James Cosmo Melvill, Robert Standen

Marine Mollusca of Madras and the Immediate Neighbourhood

Notes on a Collection of Marine Shells from Lively Island, Falklands; and...

James Cosmo Melvill, Robert Standen

Marine Mollusca of Madras and the Immediate Neighbourhood
Notes on a Collection of Marine Shells from Lively Island, Falklands; and...

ISBN/EAN: 9783744729888

Printed in Europe, USA, Canada, Australia, Japan

Cover: Foto ©berggeist007 / pixelio.de

More available books at **www.hansebooks.com**

THE MANCHESTER MUSEUM,
OWENS COLLEGE.

MUSEUM HANDBOOKS.

THE

MARINE MOLLUSCA OF MADRAS

AND THE

IMMEDIATE NEIGHBOURHOOD.

NOTES ON A COLLECTION OF

MARINE SHELLS FROM LIVELY ISLAND,

FALKLANDS;

AND OTHER PAPERS.

BY

JAMES COSMO MELVILL, M.A., F.L.S.,

AND

ROBERT STANDEN.

MANCHESTER: J. E. CORNISH.

1898.

PREFACE.

The present Handbook is the result of the study by Messrs. J. COSMO MELVILL and R. STANDEN, of collections of Marine Mollusca given to the Museum by different Friends and Collectors. The shells from Madras were obtained by Professor J. R. Henderson: those from the Falkland Islands by Miss Cobb and Mrs. Blake: whilst a number were contributed by Professor D'Arcy Thompson.

They have been reprinted from the ninth volume of the 'Journal of Conchology,' in the hope that they may prove useful to those who study the collections of mollusca in the Manchester Museum. The original pagination has been retained throughout.

WILLIAM E. HOYLE,

Keeper of the Museum.

CONTENTS.

THE MARINE MOLLUSCA OF MADRAS AND THE IMMEDIATE NEIGHBOURHOOD.

By J. COSMO MELVILL AND R. STANDEN.

(Read before the Society, Oct. 13th, 1897.)

A few years ago Professor J. R Henderson, of the Christian College, Madras, handed over the mollusca obtained during two or three dredging expeditions, in the neighbourhood of that city, to the Manchester Museum for investigation.

Want of time and pressure of other matters have, till now, prevented our accomplishing this, but we now have the pleasure of detailing the results of our examination of this very interesting collection.

As might be expected, the fauna is typically Indian, a few species showing considerable extension of range southwards, which have been till now mainly considered inhabitants of the North Indian Ocean or the Arabian Sea.

So far as we can make out, but few collections of marine shells from Madras have been formed, still fewer catalogued. That published of the contents of the Madras Museum embraces specimens from other localities as well, so that we believe the present is the first endeavour to collate such a list.

There is a large assemblage of dredged material in the British Museum, mainly collected by Mr. Edgar Thurston, Superintendent of the Madras Museum, but this has not yet been investigated.

These facts render the accompanying enumeration of greater interest than a mere list of names usually possesses.

We have thought it worth while to add to each species a note regarding its geographical distribution, and, we may remark, it is astonishing to find how very widely distributed many species are, *e.g.*, *Strombus floridus*, *S. gibberulus*, *Nerita polita*, etc. The majority of the mollusca named come from Madras and its immediately neighbouring shores, but a few were dredged in the Pamban Passage, between Port Lorne, S.E. India, and Rameswaram Island, N.W. Ceylon.

We take this opportunity of expressing our acknowledgements to Prof. Henderson for the opportunity of examining such rich and well-collected material, and we are also much indebted to Mr. Edgar A. Smith, F.Z.S., for having personally aided us in the comparison and differentiation of some obscure species; and, whilst we have left, as still doubtful, several of these, we have ventured to describe seven as new in the present paper.

The total number now catalogued comes just short of 400 species, and is therefore slightly in excess of those enumerated, three years ago, as natives of Bombay by Mr. Alexander Abercrombie and one of the present authors.[1] We should be inclined to estimate the probable total of both Madras and Bombay marine mollusca, severally, as about the same, say, 700 species or so. Both localities possess many points in common.

An asterisk (*) is appended to all those forms which are included in the Bombay catalogue just alluded to.

(I). DESCRIPTIONS OF NEW SPECIES.

Cerithium carnaticum n. sp. (Plate I., fig. 1).

C. testa attenuato-fusiformi, solida, sordidè ochracea, interdum castaneo-variegata; anfractibus novem, inæqualiter varicosis, ad suturas superficialiter canaliculatis, longitudinaliter irregulariter costatis; costis rudibus, undique transversim rudi-liratis; junctura costarum lirarumque sæpe gemmulatis; apertura ovata, labro extus effuso, paullum incrassato; canali brevi. Long. 13, *lat.* 5 *mm., sp. maj.*

It is curious that this *Cerithium* has not been described ere this, for unnamed examples exist in the British Museum. Its affinities would appear centred near *C. adenense* Sow. (which, however, is much larger) and its allies.

It is a rudely-sculptured species, solid, nine-whorled, attenuate, so impressed at the sutures as to appear channelled; the unequal varices and the irregular longitudinal ribs are crossed by thick liræ, and at the junction of these shining papillæ occur. Mouth ovate, outer lip effuse, a little thickened, canal short.

Colina selecta n. sp. (Plate I., fig. 2).

C. testa fusiformi, cylindrica, attenuata, solidiuscula, cinereo-brunnea; anfractibus undecim, apud suturas paullum impressis, undique transversim arctè sulculosis; sulcis impresso-punctatis; ultimo anfractu producto; apertura rotunda, labro exteriore effuso, incrassato, intus castaneo-lineato. Long. 15, *lat.* 4 *mm.*

Allied to *C. pinguis* A. Ad., the typical form of which is from the Cape, while varieties occur in various tropical regions, *e.g.*, Lifu and the Paumotu Is. Our species resembles more *C. tæniatum* Sow., but is not so pupiform in shape, nor is it noduled transversely. After examination of all forms of *C. pinguis* and allies, we have come to the conclusion that this is distinct from any. It is an elegant shell, and of marked peculiarity in appearance.

Rissoina (Morchiella) thaumasia n. sp. (Pl. I., fig. 3).

R. testa fusiformi, versus apicem attenuata, solida, ochraceo-alba; anfractibus novem, turritis, apud suturas paullum canaliculatis, septem superioribus profundè decussatis; costis longitudinalibus prominentibus, interstitiis quasi-punctatis; anfractu penultimo et ultimo distinctè transversim acutiliratis; costis longitudinalibus ferè evanidis; apertura obliqua; labro exteriore multum incrassato. Long. 5, *lat.* 1·50 *mm.*

---

1 *Mem. Manchester Soc.*, ser. 4, vol. 7, p. 17-51, 1893.

A beautifully sculptured *Rissoina*, allied, of course, to *R. antoni* Schwag., *R. spirata* Sow., etc., but differing from all in the decussate and strongly longitudinally ribbed sculpture of the seven upper, and in the acutely carinate transverse liræ of the two last whorls. The mouth is oblique, outer lip extremely thickened. There is one specimen in this collection and three, precisely similar, unnamed in the British Museum, also from Madras (coll. Thurston). *θαυμάσιος*, wonderful.

Syrnola maderaspatana n. sp. (Pl. I., fig. 4).

S. testa fusiformi, versus apicem multum attenuata, perlævi; anfractibus quatuordecim, apicali incluso, vitreo, pellucido, cæteris ad suturas canaliculatis, pallidissimè ochraceo-vinctis, apud suturas utrinque pellucide albo-ligatis; ultimo anfractu ad peripheriam sub lente ochracea linea succincto; apertura oblonga; labro recto, marginem apud columellarem paullum reflexo, uniplicato. Long. 10, *lat.* 2·50 *mm.*

An interesting shell, which at first gave difficulty as to precise location. Had no plait been present, we should have deemed it a *Eulimella*; it is nearer in facies to an *Obeliscus* than a *Syrnola*, but its distinct columellar plait places it in the latter genus. At first we compared it with *Obeliscus turritus* Ad., but the mouth processes are altogether different. The apex is in very perfect condition, and shows a translucent bulbosity.

Turbonilla coromandelica n. sp. (Pl. I., fig. 5).

T. testa pergracili, multum attenuata, albida, pellucida, delicatula; apice heterostropho vitreo; anfractibus quindecim, ventricosulis, undique longitudinaliter arctè recticostatis; interstitiis lævibus, nitidis; apertura trapezoide; labro extus tenui, simplice, columellarem apud marginem paullum reflexo. Long. 7, *lat.* 1·50 *mm.*

Many examples of an exceedingly graceful, attenuate, shining-white *Turbonilla*, which does not correspond with any example in the British Museum collections, nor have we seen it described or figured in any monograph. It does not approach any species nearly that we are cognizant of, the whorls being fifteen in number, delicate, pellucid, ventricosely tumid, shining, closely longitudinally straightly ribbed, the interstices being quite smooth, mouth unequally square, outer lip thin, simple, and slightly reflexed at the columellar margin.

Cadulus anguidens n. sp. (Pl. I., fig. 6).

C. testa paullum arcuata, apud apicem attenuata, pellucidè albida; apertura rotundo-ovata, margine obliquo; apertura posteriore parva, rotundo, simplici, tenui. Long. 8, *diam. oris* 1, *apicalis* 0·50 *mm.*

A graceful attenuate slightly arcuate *Cadulus*, gradually increasing in diameter till the oblique aperture is reached. The shell is subpellucid, white, quite smooth, posterior or apical orifice minute, simple, round, thin, the mouth being roundly-ovate, with very oblique margin. Two specimens, differing from any in the National collection.

Sanguinolaria hendersoni n. sp. (Pl. I., fig. 7).

S. testa tenui, ferè lævi, subnitida, obscurè concentricè inæqualiter striata; valvis posticè et anticè paullulum hiulcis; margine postico subtrapezoide, paullum producto; antico prolongato, rotundato, ventrali rectiusculo; dorsali leniter utrinque declivi; umbonibus lævibus, roseis, cætera superficie pallidè rosea. Long. 23, lat. 35 mm.

A beautiful addition to a circumscribed genus. To no known species does it nearly assimilate, save in colour, coming perhaps nearest to the West Indian *S. sanguinolenta* Gm., which, however, is far more produced and gaping posteriorly. The type, from Mr. Henderson's collection, is of the dimensions given above, but three other specimens, smaller but quite perfect (long. 20, lat. 32 mm.) exist in J. C. Melvill's collection, which were obtained at a sale at Stevens' auction rooms, in Dec., 1866, without label of locality. We have much pleasure in naming this species after its discoverer.

(II.) GENERAL CATALOGUE.

We have carefully compared the following list with that[1] compiled by Mr. Edgar Thurston, C.M.Z.S., Superintendent of the Madras Government Museum, when investigating the Zoology of Rámésvaram Island, and the Gulf of Manaar, Ceylon, and find 106 species in common. Probably the very few of Mr. J. R. Henderson's Mollusca collected at Pámban, were obtained about the same time as Mr Thurston's; the majority gathered 250 miles further north show on the whole a great dissimilarity.

Mr. Thurston's catalogue enumerates about 425 Marine Mollusca, inclusive of a few brackish water or fluviatile forms, such as *Tympanotonos*, *Potamides*, and *Melaniæ*, from Pámban and Tuticorin, which we have not mentioned though they occurred in Mr. Henderson's gatherings. Like ourselves, he has not attempted differentiation of the Chitonidæ. Our two species, both small and insignificant, are probably new, for as Mr. E. R. Sykes informs us, no Chitonidæ are yet recorded from Madras.

Amongst Mr. Thurston's more interesting records we note *Conus longurionis* Kien., which has lately occurred on the Malabar Coast (Townsend), *C. peplum* Chemn., from Muttuwartu; *Mitra zebuensis* Rv., from the same place, this being one of the finest of the genus, also *M. acupicta* Rv., *Cypræa lentiginosa* L. (also found along the whole W. Coast of Hindustan) *Pterocera scorpio* L., *Ovulum formosum* Ad. Rv., and others. We should hope that many of these will ultimately be found to reach the vicinity of Madras.

That portion of Mr. Thurston's preface which gives a glimpse of the appearance of the Madras coasts, is interesting, and well worth

1 *Bull. Mus. Madras*, No. 3, 1895.

quoting here. He says (*l.c.*, p. 79) :—"A casual observer walking along the sandy, surf-beaten beach at Madras, will probably find nothing to attract his attention excepting a number of coarse shells destined for the manufacture of chunám (lime), an occasional flattened jelly-fish, and swift-footed crabs (Ocypoda), which on the approach of man, scamper away, and disappear like rabbits into their burrows. But if the same observer walks along the shore at Pámban, he cannot help noticing that it is strewn with broken fragments of dead coral, among which branches of madrepore are most conspicuous : and sponges washed on shore by a recent tide, or dried up above water mark. And if he trusts himself upon the slimy blocks of coral which are exposed at low tide, and turns them over so as to display their under-surface, he will find there a wealth of marine life, crabs, boring anemones, annelides, shell-fish, trepangs, (bêches-de-mer), and bright-coloured encrusting sponges. And the Madras beach may, allowing for differences of species, be taken as fairly representative of the coast of the Presidency, with the exception of the coral-fringed shores of the islands which skirt the coast of the Gulf of Manaar."

CLASS GASTROPODA.

ORDER OPISTHOBRANCHIATA.

FAMILY *BULLIDÆ*.

Bulla ampulla L.—Rather small specimens. Philippines.

FAMILY *RINGICULIDÆ*.

Ringicula propinquans Hinds*—Five or six, quite normal. Philippines.

ORDER PROSOBRANCHIATA.

FAMILY *TEREBRIDÆ*.

Terebra (Euterebra) eximia Dh.—Exclusive of the new species, this is the most interesting shell in the collection. It is the second known specimen only, the type[1] in Mus. Deshayes being equally finely marked but smaller, ours measuring 48 mm. The sculpture is peculiar and very beautiful.

Terebra (Euterebra) marmorata Dh. (Pl. I., fig. 8). Many typical examples. Australia.

Terebra (Euterebra) similis E. Sm.—Described[2] from a unique individual, this being bleached. The two before us from Madras, and one from Karachi (Townsend coll.), were dredged living, are pale ochraceous yellow to fawn colour, and exhibit the characteristic sculpture. The locality having been hitherto unknown makes the discovery of these specimens of unusual interest.

---

1 Reeve, Conch. Icon., Plate xxi., fig. [illegible].
2 *Ann. and Mag. Nat. Hist.* (4) vol. 11. 1873, p. [illegible].

Terebra (Subula) crenulata L.—Several. A widely distributed species extending from the Indian Ocean to Central Polynesia.

Terebra (Subula) dimidiata L.—Common It is also found at Singapore, Philippines, and Central Polynesia.

Terebra (Subula) duplicata Lm.—In all stages of growth. A very common Indian Ocean form. Zanzibar, Madagascar, Moluccas, Singapore, China, to Fiji Islands.

Terebra (Abretia) cerithina Lm.—A few typical specimens. Also occurs from Philippines to Society Islands.

Terebra (Abretia) tenera Hinds.*—One only, but perfect and a match for Bombay and Ceylonese specimens, with which we have compared it. Ceylon and Straits of Malacca.

Terebra (Hastula) aciculina Rv.—Three specimens. The smaller size, colour, longer plications, and broad base distinguish this species from *T. cinerea* Born, with which it is often confounded. It occurs at Singapore, Manila, Marquesas, and Sandwich Islands.

Terebra (Hastula) strigilata L.—Also from Polynesia and Sandwich Islands.

Terebra (Myurella) cingulifera Lm.—Typical. Also recorded from Philippines, New Ireland, Lifu, Fiji, and China.

Terebra (Myurella) monilis Quoy.—Several. According to Tryon this is but a synonym of *T. straminea* Gray. Philippines, China.

Terebra (Myurella) myuros Lm.—Two examples of this well known species, which also occurs at Moluccas, Lifu, and New Ireland.

Terebra (Myurella) persica E. Sm.—Very beautiful, being highly chased and shining, though much grooved and latticed. A remarkable extension of its range as hitherto recognised. Persian Gulf.

Terebra (Myurella) *cf.* **turrita** Dkr.—An interesting little form, which may be distinct, but there is only one specimen, so it is hard to come to a satisfactory conclusion on the subject.

Terebra (Myurella) undulata Gr.—Also recorded from Philippines and Fiji.

FAMILY *CONIDÆ*

Conus mutabilis Chemn.*—Red Sea, St. Domingo, China.

Conus (Coronaxis) hebraeus L.—Widely distributed. East Africa, Ceylon, Mauritius, Japan, Philippines, New Caledonia to Fiji.

Conus (Coronaxis) pusillus Chemn.—A pretty little shell, which Tryon places as a variety of *C. ceylonensis* Hwass. Red Sea,

Ceylon, West Africa, Mauritius, Australia, New Caledonia, Sandwich Islands, Mazatlan, Cape St. Lucas.

Conus (Coronaxis) vermiculatus Lm. (= *C. hebraeus* L. var.)

Conus (Nubecula) gubernator Hwass.—A well known species, occurring also at Ceylon, Madagascar, Mauritius, Philippines, New Caledonia, and Seychelles Islands.

Conus (Dendroconus) betulinus L.—Two examples. The species also occurs in East Africa, Isle of Bourbon, Java, Ceylon, China, and Philippines.

Conus (Dendroconus) figulinus L.—Also from Amboyna, Java, Ceylon, New Caledonia, Lifu, and Philippines.

Conus (Dendroconus) ponderosus Beck. (= *C. quercinus* Hwass var.?)—Our specimens are old and heavy, without the revolving lines which characterise *C. quercinus*; Tryon places them together, but we have always considered them distinct. Red Sea, East Africa, Ceylon, Mauritius, Philippines, Fiji, and Sandwich Islands.

Conus (Leptoconus) amadis Martini.—Small specimens. Ceylon, Java, New Caledonia, Polynesia.

Conus (Leptoconus) generalis L.—Of ordinary character. Also recorded from Ceylon, Red Sea, Isle of Bourbon, East Africa, East Indies, Philippines, Lifu, and New Caledonia.

Conus (Rhizoconus) capitaneus L.—Quite typical. Philippines, Ceylon, Australia, Lifu, Polynesia, Mauritius.

Conus (Rhizoconus) lineatus Chemn.—A well known species, with a wide distribution. Red Sea, East Africa, Ceylon, Philippines, Australia, Lifu.

Conus (Rhizoconus) lithoglyphus Meuschen.—Also from Ceylon, Philippines, and Lifu.

Conus (Rhizoconus) magus L.—Typical. Madagascar, Borneo, Lifu, Philippines.

Conus (Rhizoconus) mustelinus Hwass.—Mauritius, Ceylon.

Conus (Rhizoconus) punctatus Sow.—Small specimens. Guinea, Ceylon, Moluccas, West Indies.

Conus (Rhizoconus) senator L.—Tryon considers this a synonym of *C. planorbis* Born, stating that the description in the 'Systema Naturæ' shows the identification of *C. senator* to be wrong, and Hanley was unable to find it in the Linnean Collection. Mauritius, Ceylon, New Caledonia, Philippines.

Conus (Rhizoconus) vexillum Gm.—Also recorded from Mauritius, Ceylon, Java, Philippines, Lifu, and Samoan Islands.

Conus (Lithoconus) virgo L.—Typical. Red Sea, East Africa, Ceylon, Philippines, New Caledonia, Polynesia.

Conus (Chelyconus) adansoni Lm.—According to Tryon, this is but a variety of *C. mediterraneus* Hwass, "Shell more cylindrical." It occurs likewise at Senegal.

Conus (Cylinder) omaria Hwass.—Typical. Red Sea, Ceylon, Philippines, Australia, Polynesia.

Conus (Hermes) nussatella L. — Small, typically marked specimens. Red Sea, East Africa, Ceylon. Java, Philippines, North Australia, Lifu, Polynesia.

Pleurotoma amicta E. Sm.* — Also common at Bombay. Described originally from the Sandwich Islands.

Pleurotoma (Surcula) tornata Dillw.- Generally known as *P. javana* L. : but that author's description is of a ribbed shell, better known as *P. nodifera* Lm., which this is not. One example, and that a fine albino. Also from Java and East Indies.

Pleurotoma (Turris) marmorata Lm.—Five, in good condition. Red Sea, Malacca, Japan, Australia, Polynesia.

Pleurotoma (Turris) variegata Kien.—Two, well marked. Indian Ocean, Japan, Philippines.

Pleurotoma (Gemmula) ceylonica E. Sm.—Characteristic, but imperfect at the mouth.

Pleurotoma (Gemmula) multiseriata E. Sm.—Two shells, both in fine condition, and better than the type. There is considerable resemblance between this and one or two of the Eocene species from Barton. Ceylon, Persian Gulf, China Sea.

Pleurotoma (Drillia) crenularis Lm. — Four, all typical. Bombay, Tranquebar, Singapore, Australia.

Pleurotoma (Drillia) major Gr.—Three specimens of a very distinct species, the habitat of which has, apparently, hitherto been unknown.

Pleurotoma (Drillia) nodifera Pease.—Many, all typical and fine. Recorded from the Sandwich Islands.

Pleurotoma (Drillia) regia Beck.—Similar to Lifu examples. Also recorded from Amboina and Moluccas.

Pleurotoma (Drillia) tayloriana Rv.—This, and *P. major* Gr., are perhaps only forms of *P. crenularis*, but are very distinct, and always recognizable. Bombay, Tranquebar, Singapore, Australia.

Pleurotoma (Clavus) echinata Lm. — Fine, quite typical. West Coast of Africa.

Mangilia horneana E. Sm.—We suspect this little Pleurotomid has a wide range, as it occurs at Bombay, though not catalogued in the last list of the molluscan fauna of that region, and Karachi also (Townsend). It has some superficial resemblance to our *M. himerodes* from Lifu,[1] but is quite distinct.

Mangilia (Cythara) cithara A. Ad. One typical example. Fiji (Gould), Paumotu Islands (Peace), Lifu (Hadfield).

Mangilia (Cythara) fusiformis Rv.—One example only, but that in superior condition: ochraceous brown, obliquely longitudinally ribbed, and transversely lirate, the interstices being much minutely and exquisitely longitudinally striate. Philippines.

Clathurella nexa Rv.—Identical with Bombay specimens. Samoa, Lifu, Philippines, Fiji Islands.

FAMILY *CANCELLARIIDÆ*.

Cancellaria (Merica) melanostoma Sow. One, very perfect. A rare form. China, Japan, Philippines.

Cancellaria (Trigonostoma) articularis Sow. (*C. scalata* Sow.). Red Sea, Ceylon, Mauritius, Moluccas, New Caledonia.

Cancellaria (Trigonostoma) crenifera Sow.—Not quite typical. Also occurs at the Philippines.

Cancellaria (Trigonostoma) scalarina Lm.* Perfect, but smaller than the ordinary form, from which they also differ in being more deeply coloured, and darker at the aperture. Mauritius.

FAMILY *OLIVIDÆ*.

Oliva (Strephona) gibbosa Lm.—Common. Ceylon, West Africa.

Oliva (Strephona) ispidula L.—In the usual variety. Philippines, Fiji and Loyalty Islands.

Oliva (Strephona) maura Lm.—Present only as var. *sepulchralis* Lm.

Oliva (Strephona) tricolor Lm.—We consider this species distinct from *O. elegans*, with which it is associated by Tryon.

Agaronia nebulosa Lm.*—In all stages of growth. Ceylon, West Africa.

Ancilla (Ancillaria) ampla Gm.—Some pretty specimens. Red Sea, Ceylon, Mauritius, Philippines.

Ancilla (Ancillaria) crassa Sow.—One poor example. Red Sea.

---

1 *J. Conch.*, vol. 5, p. [illegible], pl. [illegible], fig. [illegible].

Family *HARPIDÆ.*

Harpa conoidalis Lm.* Three: one typical, two juvenile, approaching the var. *striatula*. It is hard to distinguish where *H. conoidalis* begins and *H. articularis* Lm. ends. Ceylon, Philippines, Mauritius.

Harpa minor Rumph.—One, full grown. Indian Ocean, South Africa, Madagascar, Fiji and Loyalty Islands.

Family *MARGINELLIDÆ.*

Marginella (Volutella) angustata Sow.—Very large, the markings more or less covered by callosities: the ordinary form likewise occurring. Ceylon, Australia.

Marginella *cf.* **shoplandi** Melv.—Three, which though dead, much resemble a new species from Karachi and the Persian Gulf.[1] When in perfect condition it is transparent and exceedingly polished and shining.

Marginella (Cryptospira) quinqueplicata Lm.—Beautiful specimens. Bay of Bengal, Sumatra, Malacca.

Marginella (Gibberula) monilis L.—Two specimens. Senegal, Red Sea, Island of Socotra.

Family *VOLUTIDÆ.*

Voluta (Aulica) vespertilio L.—Small ordinary specimens. Philippines, Moluccas.

Family *MITRIDÆ.*

Mitra episcopalis L.—Several, adult and young. Ceylon, Philippines, Polynesia.

Mitra (Scabricula) crenifera Lm.—An ovate variety, prettily marked. Mauritius, Manila, Red Sea, Indian Ocean.

Mitra (Cancilla) interlirata Rv.—Four, typical. Placed by Tryon as a synonym of *M. flammea* Quoy, but sufficiently distinct, in our opinion. China, Philippines, Australia, Polynesia, Sandwich Is.

Mitra (Mitreola) litterata Lm.—A few small specimens of ordinary form. Red Sea, Java, Mauritius, South Africa, Philippines, Loyalty Islands.

Mitra (Costellaria) crebrilirata Sow.—Eight, showing some variation in size, and varying in colour from ochraceous to leaden grey. Indian Ocean, Japan, Philippines, Polynesia.

Family *FASCIOLARIIDÆ.*

Fusus longicauda Bory.—A common East Indian species, at one time confused with *F. colus* Lm. Ceylon.

1 *Manchester Mem.*, vol. 41, no. 7, p. 8, 1897.

Fusus forceps Perry. This shell which is more usually known as *F. turricula* Kien., is smaller than *F. longicauda* Bory, and [illegible] on the body whorl, the interstices being darker. China.

Fasciolaria filamentosa Lm. A prettily coloured small variety. Red Sea, Ceylon, Philippines, Australia, Loyalty Islands, Central Polynesia.

Latirus (Peristernia) pulchellus Rv. Two, small, but highly coloured, the pale pink mouth being characteristic. The columella of this *Peristernia* does not possess folds, at all events externally, thus separating it from others of the genus. Zanzibar, New Caledonia.

FAMILY *TURBINELLIDÆ*.

Cynodonta turbinella L. (= *C. cornigera* Lm.). The spines are unusually well developed. Red Sea, Moluccas, Philippines, Mauritius, Central Polynesia.

Pyrella spirillus L.—Four, one in very young state, showing a very bulbous apex. Tranquebar.

Ficula ficus L. (= *F. lævigatus* Rv.).—Many, all the ordinary form. Red Sea, Indian Ocean, Singapore.

Ficula reticulata Lm. — Typical examples. Sooloo Archipelago, Indian Ocean, East Indies, Japan.

Rapella (Melongena) paradisiaca Rv. Showing some variation. Ceylon, Red Sea, Mozambique, Natal.

Rapella (Melongena) pugilina Born. Quite young, but unmistakeable. Indian Ocean.

Hemifusus lacteus Rv. Two of this rare form. Philippines.

FAMILY *BUCCINIDÆ*.

Cyllene fuscata A. Ad.*—Also from Malacca and Singapore. According to Tryon this is a synonym of *C. lugubris* Ad. Rv.

Pisania spiralis Gr.* Several. Mauritius.

Tritonidea tranquebarica Gm.—Several specimens.

Tritonidea undosa L.—Fine, with operculum. Malacca, Australia, Philippines, Fiji, Paumotus, and Loyalty Islands.

Engina armillata Rv.—A very fine shell, in a beautiful state of preservation. Philippines, Arakan.

Engina pulchra Rv.—One, of the pale white wreathed variety. Panama, Galapagos, and Loyalty Islands.

Nassaria suturalis A. Ad.* (= *N. acuminata* Rv.).—Five: this *Nassaria* seems generally, but sparsely, distributed from Karachi through Bombay and Ratnagiri to the Laccadives and Ceylon, and now found in Madras. China Sea, Indian Ocean.

Latrunculus[1] spiratus Lm.* Several, one operculated. Ceylon, Philippines.

Latrunculus zeylanica Brug. Several, of ordinary form Ceylon.

FAMILY *NASSIDÆ.*

Nassa (Arcularia) callosa A. Ad.—Very finely grown, showing the white callosity at the mouth. Philippines, Indian Ocean, Lifu.

Nassa (Arcularia) labecula A. Ad. (= *N. jonasi* Dk.).—This may be only a variety of the preceding species, from which it is distinguished principally by form and sculpture. Port Jackson, Australia.

Nassa (Arcularia) leptospira A. Ad.—Quite typical. Philippines (Cuming), Japan, Ascension Island (Pease).

Nassa (Alectryon) monile Kien.—A very handsome variety. Philippines, Australia, Central Polynesia, Lifu (Hadfield).

Nassa (Alectryon) mucronata A. Ad. (= *N. monile* Kien. var.*) Very fine, in better condition than our Bombay examples. Loyalty Islands, Australia, Philippines.

Nassa (Alectryon) scalaris A. Ad.—Our specimens possess some of the characters of *N. monile* and *N. papillosa*. Philippines (Cuming).

Nassa (Niotha) splendidula Dkr.—Only one of this choice shell, which appears almost typical. A brightly banded and highly sculptured *Nassa*, in shape like a small *N. stigmaria*. Philippines, Malacca, Lifu, Polynesia.

Nassa (Niotha) stigmaria A. Ad. The typical form, which comes near *N. reticosa* A. Ad., *N. candens* Hinds, and *N. cremata* Hinds, all high-class *Nassæ*, with elaborate sculpture. Varieties of this species occur throughout the Indian Seas, and are especially frequent in the Andaman Isles. Philippines, Malacca.

Nassa (Hima) plebecula A. Gd. — One, undoubtedly this species, which Tryon unites with *N. paupera* A. Gd. Japan, Australia, Lifu (Hadfield).

Nassa (Hima) stolata Gm.—Many, varying in size and form. Some are elongate, others ventricose and more robust, shewing the same form that occurs so plentifully on the western and southern shores of India. More generally known by the now superseded name *ornata* Kien.

Nassa (Zeuxis) canaliculata Lm. — Many, very handsome, well developed, and rich in colour, varying from orange-brown to grey. Philippines, Polynesia.

1 It is with regret that the familiar name *Eburna* Lm. can no longer be used, having been employed twice by that author—firstly, as a synonym of *Ancilla* Lm. in 1801, and subsequently (1[illegible]) applied to the old *Buccinum spiratum* L., and its allies.

Bullia (Dorsanum) belangeri Kien.* Many very fine living examples. Arakan, Ceylon.

Bullia (Dorsanum) cumingiana Dkr. Two, evidently of this species, the habitat of which is unrecorded.

Bullia (Dorsanum) livida Rv. Two, fine examples.

Bullia (Dorsanum) vittata L. Many, some albino, fine living shells. Ceylon, Zanzibar.

FAMILY *COLUMBELLIDÆ.*

Columbella (Mitrella) euterpe Melv.*—Many examples, agreeing precisely with the original description, the types having been unfortunately mislaid.

Columbella (Mitrella) flavilinea Melv.* Three specimens.

Columbella (Nitidella) ala-perdicis Rv. (= *C. lævigata* L. var.) Two, in young condition. West Indies.

Columbella (Anachis) terpsichore Leathes.* Found both in the Eastern and Western Hemisphere. One specimen, but large and in fine condition. West Indies.

Columbella (Pygmæa) flavida Lm.—Synonymous with *C. flava* Brug. Indian Ocean, Japan, Mauritius, Seychelles, New Caledonia, Polynesia.

Columbella (Pygmæa) versicolor Sow.—A variable species, of wide distribution, and with—according to Tryon—a long array of synonyms. The oldest name for this species is *C. scripta* Lm., but Linnæus had previously used this name for a well-known Mediterranean species. Indian Ocean, Japan, Philippines, Australia, Polynesia.

FAMILY *MURICIDÆ.*

Murex brevispina Lm.—Quite typical. Red Sea, Indian Ocean, South Africa, North Australia.

Murex tribulus L.* Several. Owing to the shortness of the spines, and the variable transverse nodulosities on the last whorls, we should consider our specimens as coming under the var. or sub-species *M. tenuispina* Lm. We cannot see our way to allow true specific rank to this ancient species.

Murex (Haustellum) haustellum L.—One only, but fine. Ceylon, Indian Seas, Red Sea, China, Mauritius, Philippines.

Murex (Chicoreus) microphyllus Lm.—Large, and in fine condition. Ceylon, Indian Ocean.

Murex (Ocinebra) contractus Rv. Exactly corresponding with Bombay examples. New Caledonia, Philippines, Fiji Islands.

Murex (Phyllonotus) anguliferus Lm.—In all stages of growth. Red Sea, Indian Ocean, Seychelles, Isle of Reunion.

Rapana bezoar L. One specimen, of the typical ribbed form, in fine condition. China, Japan, Philippines.

Rapana bulbosa Soland. Many, very fine, and in all stages of growth, with opercula. China, Japan, Philippines.

Purpura rudolphi Lm.- Hardly typical, and somewhat juvenile, but strongly filletted and well marked. Philippines.

Purpura tissoti Petit.* — Identical with Bombay examples. The late Mr. Tryon's remarks as to this species [1] prove that he did not know it, and that Mr. Swift, whom he quotes, was also misled into considering it a var. of the West Indian *Cantharus coromandelianus* Lm.

Purpura (Stramonita) bufo Lm.*—Many specimens in every stage of growth, with opercula, and showing some variety. Philippines.

Purpura (Polytropa) sacellum Chemn.*—Also common at Bombay. Many, both young and mature.

Purpura (Cronia) amygdala Kien.—Typical; the close ribs thickly covered over with small arched scales. Australia.

Sistrum iostoma Rv. Quite typical. The habitat of this species is hitherto unrecorded.

Sistrum konkanense Melv.*—Described as a *Ricinula*, but *Sistrum* has priority. The range of this species, it is interesting to observe, is being gradually extended both north to Karachi, and south to Ceylon, and eastward. The example before us is hardly typical, being not so elongate as the Bombay shells.

Sistrum margariticolum Brod.- Several of a shell which has not been fully understood by students till recently, but which, in our opinion, is a good species. Mauritius, Loyalty Islands, Australia.

Sistrum tuberculatum Blainv.*- Common and very variable. The animal is described by Gould in the "Mollusca" of the Wilkes' Expedition as being deep grass-green, with the mantle, locomotive disc, and tentacles light sea-green, finely dotted with white. Japan, Philippines, to Sandwich Islands.

Sistrum undatum Chemn.—Several typical examples. Tranquebar, Natal, China, Japan, Australia, Polynesia to Paumotu Islands.

FAMILY *TRITONIDÆ.*

Aquillus chemnitzi Gr. (=*A. tranquebaricus* Lm.) – Some small specimens. Panama, West Coast of Africa, West Indies. We consider *Aquillus* has precedence of other names proposed for this genus, and is not too like *Aquila* in ornithology to be discarded in consequence of such similarity; doubtless the derivation was from 'aqua,' water.

[1] Man. Conch., vol. 2, p. 164, under *Cantharus*.

Aquillus (Lotorium) lotorium L.—A common Ceylonese species. Red Sea, Indian Ocean, Philippines, Central Polynesia.

Aquillus (Simpulum) aquatilis Rv.*—One, exceeding [illegible] and well marked. Loyalty Islands.

Aquillus (Lotorium) retusus Lm.—Some typical specimens. Indian Ocean, Mauritius.

Aquillus (Lotorium) tripus Chemn.—Small, well marked specimens. Indian Ocean, China.

Distortrix cancellinus Roissy.—One specimen. Ceylon, China, Philippines, St. Thomas and other West Indian Islands, Monte Christi and Xipixapi, West Columbia.

Gyrineum[1] **crumena** Lm.—Many examples, of ordinary form. Ceylon, Philippines.

Gyrineum (Bursa) margaritula Dh.—Typical examples. Indian Ocean.

Gyrineum (Bursa) spinosa Lm.*—A large number of specimens, in all stages of growth, with opercula. Red Sea, Indian Ocean, Philippines, Mauritius.

Gyrineum (Lampas) affinis Brod.—Several, quite typical. Philippines, Loyalty Islands, Samoa, West Indies.

Gyrineum (Apollon) tuberculata Brod.*—Medium sized specimens. Indian Ocean, Red Sea, China, Malacca, Manila, Tahiti.

FAMILY *CASSIDIDÆ.*

Cassis (Semicassis) canaliculata Lm.—Several examples. Ceylon, Philippines.

Cassis (Semicassis) sulcosa Brug.—Not quite typical: our examples agreeing very well with *C. undulata* Gm., which Tryon considers a variety only. Mediterranean, Portugal, West Coast of Africa, West Indies, Brazil, West Coast of North America from Panama to Guaymas.

Cassis (Phalium) areola L.—Two good specimens. Indian Ocean, Malacca, Philippines, New South Wales.

FAMILY *DOLIIDÆ.*

Dolium chinense Dillw. (=*D. variegatum* Lm. var.) China, North Australia.

Dolium costatum Mke.—Of ordinary form. East Indies, Philippines, Mauritius.

Dolium fasciatum Brug.—One only, with unusually fine varix. Philippines, China, Japan.

---

1 The well-known name *Ranella* Lm. (1812) is superseded by *Gyrineum* Link (1807). *Bufo* Montfort (1810).

Dolium fimbriatum Sow. (= *D. costatum* Mke. var.)—Several small specimens. Senegal, Manila, Japan.

Dolium perdix L.—One specimen of this widely-distributed species. Indian Ocean, Mauritius, Lifu, Polynesia, West Africa, West Indies, Brazil.

FAMILY *CYPRÆIDÆ*.

Simnia sowerbyana Weink.—One, in fine condition, of pale-flesh colour, transversely indistinctly banded with white. This is *Ovula spelta* Sow. & Rv., not Lm. South Seas.

Cypræa (Aricia) annulus L.*

Cypræa (Aricia) arabica L.*

Cypræa (Aricia) caput-serpentis L.

Cypræa (Aricia) moneta L.*

Cypræa (Aricia) reticulata Martyn.—A medium-sized form of var. *intermedia* Gr.

Cypræa (Luponia) errones L.

Cypræa (Luponia) lynx L.

Cypræa (Luponia) ocellata L.*

Cypræa (Luponia) pallida Gr.*—Extremely fine.

Cypræa (Luponia) vitellus L. With few exceptions, all the above common species are of the most ordinary character, and call for no special comment; they are of the usual types met with on the Indian coasts. Nearly all are widely distributed over the whole Indo-Pacific region.

Erato pellucida Rv.*—A widely-distributed species of *Erato*; we have seen it from the shores of Beluchistan, also Karachi, Bombay, and Ratnagiri.

FAMILY *STROMBIDÆ*.

Strombus lentiginosus L.—Typical. Philippines, Zanzibar, New Caledonia to Fiji.

Strombus (Canarium) urceus L.—Two good specimens, with dark orange-coloured mouth. Indian Ocean, Philippines, Australia.

Strombus (Canarium) gibberulus L.* — Normally-marked examples of this widely-distributed species. Zanzibar, Natal, Red Sea, Indian Ocean, Mauritius, Philippines, Lifu, New Guinea, Fiji and Paumotu Islands.

Strombus (Gallinula) marginatus L.—Several pretty specimens. Formosa, China, Philippines.

Strombus (Gallinula) isabella Lm.—Three, shewing we think peculiarities specific from *S. canarium* L., with which this shell has been united by some authors. The mouth is wider, and the oute lip more effuse and angular, and not so incrassate; colour a pale unicolorous isabelline yellow, which no doubt suggested to Lamarck the trivial name.

Strombus (Conomurex) luhuanus L. Three, quite typical. Australia, New Guinea, Philippines, Loyalty and Fiji Islands.

Pterocera (Harpago) chiragra L. —Only in young state. Indian Ocean, Philippines, Loyalty Islands, Polynesia.

Rostellaria curvirostris Lm. Several, mostly in a young state. Red Sea, Moluccas.

FAMILY *CERITHIIDÆ*.

Cerithium carnaticum M. & S., *vide antea*, p. 31.

Cerithium *cf.* **corallinum** Defr.—Three examples, which we consider very near to, if not identical with, this species.

Cerithium litteratum Born.—Several specimens. West Indies, Florida.

Cerithium morus Lm.—A well-known, variable, and widely-distributed species. Madagascar, Red Sea, Philippines, Australia, Loyalty and Fiji Islands.

Cerithium nodulosum Brug.—One fine specimen. Singapore, Moluccas, Philippines, Lifu (Hadfield).

Cerithium vulgatum Brug.—Some typical examples. There are few shells with such an enormous synonymy of varietal names as this species, although it does not vary more than is usual in the genus. Tryon figures a number of these so-called varieties, but does not recognise their claim to separation from the type. Southern Europe, West Africa, Cape of Good Hope.

Cerithium yerburyi E. Sm.—Allied to *C. morus* L. but more attenuate. Originally described from Aden, but no doubt it will be found all along the Indian Coasts.

Colina macrostoma Hinds.—A very interesting form, of which only one occurred in Prof. Henderson's dredgings. Straits of Malacca, Borneo.

Colina selecta M. & S., *vide antea*, p. 31.

Pyrazus palustris L.—Two, of typical form. India, Java, Australia, Loyalty Islands.

FAMILY *PLANAXIDÆ*.

Planaxis nigra Quoy.—Six, of a plain-coloured, smooth, unpolished shell. New Ireland, Fiji and Sandwich Islands, South Africa.

Planaxis sulcatus L.*—Eighteen, all well grown and unusually large. Indian Ocean, Sandwich Islands, Australia, Philippines, Mauritius, South Africa.

FAMILY *TURRITELLIDÆ*.

Turritella triplicata Stud.—Three, of normal coloration. Mediterranean, West Africa, Canary Islands, Coast of Spain.

Turritella (Haustator) columnaris Kien.—One, in poor condition. Ceylon.

Turritella (Zaria) duplicata L.*—Many, including some perfect and large albino varities; also var. *β. attenuata* Rv. is present in all stages of growth. Indian Ocean generally.

FAMILY *LITTORINIDÆ*.

Littorina scabra L.—Plentiful, of typical form. Including its varieties, this species stretches nearly round the world, extending from West Africa around to Arabia, but has not yet been found in the Mediterranean Sea. Indian Ocean, China, Fiji, Sandwich, and Philippine Islands, Mazatlan, Florida, Ceylon, Japan, West Coast of Africa, West Indies, Polynesia.

FAMILY *SOLARIIDÆ*.

Solarium delectabile Melv.* — Described in 1893 from Bombay specimens collected by Mr. Abercrombie. Our Madras specimen is unmistakable, though not in prime condition. This discovery extends the range of this very beautiful little *Solarium*, which will probably be found before long distributed around the Indian coasts.

Solarium modestum Phil.—Two pretty specimens. This may, perhaps, be a variety of *S. perspectivum*, from which it differs in the colouring only. Society Islands, China.

Solarium perspectivum L.—One specimen is more conical than usual. It also occurs in *statu juvenili*. Amboyna, Indian and Pacific Oceans, China, to Australia.

Solarium pictum Phil.—One, rather worn. New Guinea.

FAMILY *RISSOIIDÆ*.

Rissoina (Phosinella) deshayesi Schwartz.—With some little diffidence we name our solitary large clathrate *Rissoina* as above. Singapore (in Mus. Brit. unnamed), Philippines.

Rissoina (Phosinella) sp. — To some extent agreeing with Lifu examples of *R. quasillus* M. & S., but coarser in texture. We also have the same form from Thursday Island, sent in shell sand, collected by Mr. Arnold Henn.

Rissoina (Morchiella) thaumasia M. & S., *vide antea*, p. 31.

Fenella cerithina Phil*.—Two, rather worn. A small, prettily decussated species. Mauritius, Island of Rodriguez, Red Sea, Persian Gulf, Japan.

Iravadia trochlearis A. Gld.*—Many, in excellent condition. Evidently abundant throughout the Indian Ocean. We have many specimens from Bombay, and have seen it from Ceylon, Karachi, and Persian Gulf. Japan, Hong Kong.

FAMILY *CAPULIDÆ.*

Amathina tricostata Gm.—Very perfect, though not large. Only one example. East Indies, Japan.

Crucibulum (Dispotæa) extinctorium Lm. (=*C. scutellatum* Gr. var.)—Three, fine. West Indies, Ceylon, West Coast of America from Chili to Mazatlan.

Crepidula (Crypta) scabies Rv.—Many specimens, in good condition. Ceylon, Singapore, China Sea, Japan.

Crepidula (Ergæa) walshi Herm. (=*C. plana* Ad. Rv. var).—Many good and characteristic examples. Abundant also at Bombay. Japan, Singapore, China, Ceylon.

Calyptræa diaphana Rv. (=*Mitrularia equestris* L. var).—Many examples. Tryon has "interpreted this species in accordance with general usage, the Linnæan species being indeterminable." He gives a long synonymic list of forms representing such variety in shape and sculpture, that it is not surprising that they were described as distinct species by the older school of conchologists. China Sea, Philippines, Indian Ocean, Prince's Island, West Africa, West Indies, West Coast of Central America, Galapagos Islands.

Calyptræa fibulata Rv. (=*Mitrularia equestris* L. var.)—Several good examples. Philippines, West Indies.

FAMILY *XENOPHORIDÆ.*

Xenophora solaris L.—One beautiful example. Malacca, Singapore.

FAMILY *LAMELLARIIDÆ.*

Lamellaria perspicua L.—Neither we nor Mr. Edgar Smith can discover any means of differentiating the Madras form from that found in English waters.

FAMILY *NATICIDÆ.*

Natica lineata Lm.*—Medium-sized specimens, of typical form. Singapore, Philippines.

Natica marochiensis Lm. (=*N. marocana* Chemn.)—The many examples in this collection of a small cinereous shell, quite plain, and concentrically wrinkled in a characteristic manner round the sutures, we cannot exactly identify, but presume it is one of the many forms of Lamarck's protean species. West Africa, West Indies, Panama to Mazatlan, Society and Philippine Islands, Lifu, Australia.

(*To be continued*).

THE MARINE MOLLUSCA OF MADRAS AND THE IMMEDIATE NEIGHBOURHOOD.

By J. COSMO MELVILL and R. STANDEN.

(Read before the Society, Oct. 13th, 1897).

(*Continued from page 48*).

Natica pulicaria Phil*. One small individual, exactly agreeing in marking with a large specimen so named in J. C. Melvill's collection. Habitat, hitherto unknown.

Natica (Neverita) chemnitzi Récluz (=*N. ampla* Phil. var.).—Three typical examples. Indian Ocean, China, Japan, Australia, Mauritius.

Natica (Polinices) columnaris Récluz.—Several small specimens. Philippines, Mauritius.

Natica (Polinices) mamilla L.—The numerous specimens we have are medium-sized, and typical. East Indies, Lifu, Central Polynesia, Philippines.

Natica (Ruma) zanzibarica Récluz (=*N. melanostoma* Gm. var.).*—Our specimens are a little more quadrangular than the type of *N. melanostoma*. East Indies, Mauritius, Madagascar, Philippines, Western Polynesia.

Sigaretus javanicus Gr.—Some good examples. East Indies.

Sigaretus neritoides L. Several examples of this common East Indian *Sigaretus*, in a very good state of preservation. Some confusion attends the limitation of the various forms of this genus, and Tryon considers *S. javanicus* and *S. neritoides* identical.[1] We can hardly agree with him.

Sigaretus (Catinus) planulatus Lm.*—Several; easily distinguished by its smooth, flattened surface. Australia, Philippines, Zanzibar.

FAMILY *IANTHINIDÆ*.

Ianthina globosa Sw.—A beautiful and perfect example, pale violet in colour.

FAMILY *SCALARIIDÆ*.

Scalaria (Scala) tenuicostata Sow.—Quite perfect and very beautiful, the ribs being very closely set. Japan.

Scaliola bella A. Ad.—One, seemingly identical with this Japanese species. Our example is only eight whorled, however, as against nine. It may very likely be an undescribed form, of which more material is wanted before deciding.

Aclis eoa Melvill.*—One, agreeing with the type from Bombay.

1 Man. Moll., vol. 8, p. 55.

FAMILY *EULIMIDÆ.*

Apicalia holdsworthi H. Ad.—Seven specimens of this interesting Stylifer. Ceylon.

FAMILY *PYRAMIDELLIDÆ.*

Obeliscus pulchellus A. Ad.* Japan, Loyalty Islands (Hadfield).

Obeliscus terebellum A. Ad.—Very perfect. Antilles.

Syrnola maderaspatana M. & S., *vide antea*, p. 32.

Pyrgulina interstriata Sow.—A few, in poor condition, seemingly identical with Bombay and Upolu examples in J. C. Melvill's collection.

Pyrgulina kreffti Angas.—One, of what may be this Australian form.

Turbonilla candida (Ad.) One fine shell, in all respects agreeing with examples dredged by Mr. F. W. Townsend in the Arabian Sea (Karachi, etc.)

Turbonilla coromandelica M. & S., *vide antea*, p. 32.

Cingulina spina Cr. Fisch.—Many examples of an elegant species.

FAMILY *NERITIDÆ.*

Nerita crassilabrum E. Sm.—Several. Red Sea, Indian Ocean, Natal, Singapore, China, Philippines, Fiji Islands.

Nerita gemmulata Rv.—Three specimens, agreeing well with description and examples in British Museum. Habitat hitherto unknown.

Nerita haustrum Rv.—A black Nerite, finely transversely sulcate. Tryon considers this synonymous with *N. yoldi* Récl. which, like some other West Indian species is subject to an erosion which excavates the white portions, leaving the black in relief. Red Sea, Indian Ocean, Hong Kong.

Nerita histrio L.—Some handsome, well-grown individuals in very fine condition. Australia, Mauritius, East Africa, East Indies, Philippines, Polynesia.

Nerita polita L.*—A number of examples, showing the usual variation in colour so notable in this common but beautiful species. Red Sea, Indian Ocean, Philippines, Mauritius, Loyalty Islands, Polynesia.

Nerita (Thelicostyla) albicilla L.*—Several of this widely-distributed species. Natal, Singapore, China, Philippines, Loyalty and Fiji Islands, Red Sea, Indian Ocean, &c.

Nerita (Peloronta) plicata L.—Some typical specimens. Formosa, Indian Ocean, Polynesia, Loyalty and Sandwich Islands.

Neritina mertoniana Récl.—Many specimens. This is probably, as considered by Tryon, a variety of *N. natalensis* Less., which closely mimics the common West Indian *N. virginea*, L., and is equally variable in its markings. Indian Ocean to Philippines and Polynesia.

FAMILY *TURBINIDÆ.*

Phasianella (Orthomesus) variegata Lm.—We give the above name to two examples of a small, smooth, closely interruptedly white-lined *Phasianella*, with some little doubt, the exact variety not being found in the Mus. Brit., but it seems near *P. nivosa*, *P. lentiginosa*, and other quasi species now aggregated by Pilsbry under the above name.[2]

Turbo (Senectus) radiatus Gm.—Many, in all stages. Red Sea to Madagascar, eastward to New Caledonia, Nicobar, Philippine, and Loyalty Islands.

FAMILY *TROCHIDÆ.*

Polydonta maculata L. var.—Several specimens of this protean species, differing from the type in some respects. Philippines, Singapore, Fiji Islands, Indian Ocean, Kingsmill Island.

Polydonta veneta Rv.—A few examples of a species allied to *P. radiata* Gm. but without teeth on the columella. Moluccas (Rv.).

Polydonta (Carinidea) radiata Gm.*—Many, in all stages of growth. Red Sea, Indian Ocean, Singapore, Madagascar, Ceylon.

Umbonium vestiarium L.*—A large number of specimens. apparently not so variable or light in colour as examples from the Western shores of India. The var. *rosea*, however, is present, and the variety with slaty black umbilical callosity. Indian Ocean, Ceylon to Java, Philippines, Singapore, New Irealand.

Gibbula nuclea Phil.—Typical specimens. New Caledonian Archipelago, Japan, Fiji Islands (Garrett).

Minolia biangulosa A. Ad.—An abundant South Indian *Minolia*, the whorls being excessively angulate. We have lately seen dredged from Ceylonese waters by Capt. Tindall, of the s.s. "Patrick Stewart," over a hundred, showing no variation whatever. Siam.

Minolia variabilis A. Ad.—Also extremely abundant, and, as its name would imply, variable. Persian Gulf.

Calliostoma interruptum Wd.—We have identified this by Reeve's figure and description, not having seen any named individuals. It is an elegant little trochiform shell, prettily longitudinally banded, and articulately filletted at the periphery. Habitat hitherto unknown.

2 Tryon, Man. Conch. vol. 10, p. 179.

Calliostoma tranquebaricum Chemn. Many, all fine, and exhibiting little variation in marking, none in sculpture. Tranquebar, Pondicherry, Vizagapatam.

Euchelus atratus Gm.—Darker in colour than those in the British Museum. Nicobar, Fiji, and Philippine Islands, Vanikoro, Moluccas, Sorong, Singapore.

Euchelus tricingulatus A. Ad.—Several of a pretty fawn-coloured shell. Malacca, Singapore.

Euchelus foveolatus A. Ad.—One or two specimens of a very distinct little shell, with coarsely latticed sculpture. Philippines, Lord Hood's Island, Paumotu, and Fiji Islands.

Euchelus horridus Phil.* Rather small and poor, examples of an abundant Eastern shell. Bombay, Mediterranean.

Euchelus indicus A. Ad.*—Large and well marked, and in good quantity. Bombay.

Euchelus proximus A. Ad.*—Only one, but fine. Tryon considers this a form of *Trochus asper* Gm., though, typically, this is more elevated, with smaller and more numerous spirals. Indian Ocean.

Euchelus scaber P. Fisch.—Three typical specimens. Indian Ocean, Singapore (Archer).

Euchelus tricarinatus Lm. (=*E. quadricarinatus* Chemn. var.)*—Several; Indian Ocean.

FAMILY *DELPHINULIDÆ*.

Liotia varicosa Rv.—Occurs also at the Philippines and Loyalty Islands (Hadfield).

FAMILY *CYCLOSTREMATIDÆ*.

Cyclostrema pulchellum Dkr.—Identical with Japanese examples. Singapore, Australia.

FAMILY *HALIOTIDÆ*.

Haliotis (Teinotis) asinina L.—Several. China, Japan, Australia, East Indies.

Haliotis glabra Chemn.—Several; a smooth species, readily distinguished by its oval form and the green arrow-shaped blotches on the surface. Philippines, Australia.

Haliotis varia L.—Three examples: a well-named species, of wide distribution. Philippines and Australia to China, Mozambique, Red Sea, Mauritius, Ceylon, Nicobar Islands, Malay Archipelago.

FAMILY *FISSURELLIDÆ*.

Fissurella bombayana Sow.*—Two; quite typical.

Fissurella (Glyphis) lima Sow. Quite typical. Bombay, Arakan.

Fissurella (Glyphis) ruppelli Sow.—Some characteristic examples. Mauritius, Red Sea, Cape of Good Hope.

Emarginula costulata Dh.—One Madras example, and three from Bombay (Abercrombie) hitherto unnamed in J. C. Melvill's collection, with some slight doubt we refer as above. Id. of Réunion.

Subemarginula (Clypidina) notata L.*—Very prettily marked with slatey indigo concentric lines and dashes. Tryon remarks that the habitat, West Indies, given by Adams is doubtful, and that it belongs rather to an East Indian type, an opinion borne out by this record of ours from Madras, as well as a former one from Bombay.

Scutus *cf.* **corrugatus** Rv.—Only small forms. Japan.

FAMILY *ACMÆIDÆ.*

Acmæa saccharina L.—Some fairly typical specimens. Amboyna to Japan; Fiji Islands.

FAMILY *PATELLIDÆ.*

Helcioniscus variegatus Rv. (= *H. rota* Gm. var.).—Many specimens, agreeing with Tryon's definition. Suez and Mozambique (Rv.); Id. of Réunion; Madagascar (Dall).

CLASS POLYPLACOPHORA.

FAMILY *CHITONIDÆ.*

Ischnochiton sp.—Two small species which have not yet been worked out, but both are probably new to science. We shall hope to say more about them in a subsequent paper.

CLASS SCAPHOPODA.

FAMILY *DENTALIIDÆ.*

Dentalium octogonum Dh.—Several specimens.

Antalis fissura Lm.—One perfect example of this rare species.

Cadulus anguidens M. & S., *vide antea*, p. 32.

CLASS PELECYPODA.

ORDER TETRABRANCHIATA.

FAMILY *OSTREIDÆ.*

Ostrea denselamellosa Lischke.—Two characteristic examples, agreeing with Japanese specimens in the British Museum, and J. C. Melvill's collection.

FAMILY *ANOMIIDÆ.*

Anomia humphreysiana Rv.—Only one valve, but characteristic.

FAMILY *LIMIDÆ.*

Lima squamosa Lm.—Several, approaching the typical Red Sea form. Red Sea, Mediterranean.

Lima (Mantellum) angulata Sow.—Several, resembling exactly specimens in the British Museum from Panama. Bay of Caraccas.

FAMILY *PECTINIDÆ.*

Amussium japonicum Gm. Quite typical. A series of adult and young. China, Japan.

Amussium pleuronectes L. Typical examples, in all stages of growth. China.

Pecten asper Sow.—This is one of the most interesting of the recent Pectinidæ. The markings are peculiar, as is the sculpture, and well represented in plate i. of Reeve's Conch. Icon. New Guinea (Hinds).

Pecten crassicostatus Sow. Several typical examples. Japan.

Pecten cristularis Ad. & Rv.—Some pretty specimens. Distributed throughout the Indian Ocean.

Pecten singaporinus Sow.*—Only one, a finely coloured but small specimen.

Pecten sinensis Sow.—Several examples, in various stages. China.

Pecten squamatus Gm.—Typical examples of this inequilateral Philippine Island species.

Pecten tranquebaricus Lm.—Small but perfect, and highly characteristic, shewing the peculiar acute auricles, and of a pale ochraceous colour. Coasts of Tranquebar.

Pecten (Chlamys) limatulus Rv. Three specimens of this delicate and elaborately-sculptured species. Mauritius.

FAMILY *AVICULIDÆ.*

Avicula argentea Rv.—A smooth, dark olive shell. Also reported from the coasts of Guinea.

Avicula formosa Rv.—Through a lens, this shell, much encrusted as it is with nullipores, etc., is seen to be most delicately concentrically lirate.

Avicula iridescens Rv.—Two specimens. Moluccas.

Avicula scabriuscula Rv.—Two; characteristic. Australia.

Margaritifera anomoides Rv.—Many; a decided species, semitransparent, white, rayed with pale green in a manner very unusual in this genus. Philippines.

Margaritifera margaritifera L.—Medium and juvenile examples of this abundant tropical species.

Margaritifera prætexta Rv.—Many; of a peculiar livid tawn colour, the pale oblong blotches with which the shell is rayed are curiously wrinkled across with opaque-white lines. Philippines.

Margaritifera squamulosa Lm.—Young examples, which are probably juvenile forms of *M. flabellum* Rv.; if so, the Lamarckian name, having priority, must stand. They are beautifully concentrically squamate, the squamæ long and very fragile. Tiger Bay.

Margaritifera tegulata Rv.—Very delicate and beautiful Roundish, thin, and thrice or four times longitudinally radiate. We also have the young form, pale green, and very oblique, with the characteristic radiate lines. Also from Moreton Bay.

Margaritifera vexillum Rv.—Many, of a prettily zig-zag marked Avicula, but all in quite young condition. Ceylon.

Perna femoralis Lm.—Several, quite typical. Philippines.

Perna isognomon L.—A few of this common East Indian form. Philippines.

Pinna attenuata Rv.—Two specimens; typical. Moluccas.

Family *MYTILIDÆ*.

Mytilus viridis L.—Small, but well coloured.

Modiolus metcalfei Hanl.—A delicate species, very perfect in condition. Philippines.

Modiolus ramosus Hanl.—Two specimens. Celebes.

Lithophagus stramineus Dk.—Two. West Indies.

Lithophagus teres Phil.—Three small sized specimens. A shell of simple form and sculpture. Mazatlan.

Modiolaria cœnobita Vaill.—An interesting form, described by Vaillant from the Gulf of Suez.

Family *ARCIDÆ*.

Arca imbricata Poli.—Very similar to the European *A. tetragona* L. Aden (Brit. Mus.), Mediterranean, East Indies.

Arca inæquivalvis Brug.*—Many, in all stages. Indian Ocean, Persian Gulf.

Barbatia fusca Brug.—A quantity of this common Philippine species.

Barbatia lima Rv.—Two or three only. Philippines.

Barbatia (Acar) pusilla Sow.—A neat squarrose shell, identical with Tasmanian specimens received by the British Museum from Mr. J. H. Ponsonby.

Barbatia (Venusta) lactea L.*—Of very wide distribution, being a common British and European species. Also from Bombay and Arakan in the Blanford collection (Mus. Brit.). Mediterranean.

Scapharca clathrata Rv.—Several. Philippines (Cuming), Aden (Brit. Mus.).

Scapharca rhombea Born.*—A solid handsome shell. China, Ceylon.

Parallelipipedum tortuosum L. Three specimens of this extraordinary shell. Singapore, Malacca.

Cucullæa concamerata Chemn.—Two, very fine. Indian Ocean (Paetel).

Pectunculus taylori Angas.—Two odd valves, both quite distinctive and full grown. The geographical distribution includes Ceylon.

FAMILY *NUCULIDÆ*.

Nucula mitralis Hinds. A glossy and very oblique small species. Also from Malacca.

Nuculana cuspidata A. Gd.—Many beautiful examples, absolutely identical with a solitary specimen, in the British Museum, with no locality. Gould's types came from North America, and it is strange to have it reported from Madras.

FAMILY *CARDITIDÆ*.

Cardita canaliculata Rv.—One example only, and that in juvenile condition. Those in the British Museum are from the Philippines.

FAMILY *TRIDACNIDÆ*.

Tridacna gigas L.—One small specimen in a worn condition. Philippines.

FAMILY *CARDIIDÆ*.

Cardium (Acanthocardia) asiaticum Chemn.—Several, in various stages. China, Nicobar Islands.

Cardium (Trachycardium) rugosum Lm. (= *C. flavum* L.). Several. Madagascar, Ceylon, Nicobar Islands.

Cardium (Cerastoderma) latum Born.*—Some characteristic examples. Philippines (Cuming).

Cardium (Bucardium) coronatum Spengl.—Typical specimens. China.

Cardium (Bucardium) fimbriatum Wood.—Many examples. China.

Cardissa (Lunulicardia) subretusum Sow.—Several. These are a small form of what Paetel and others consider a variety of *C. retusum* L. Red Sea.

FAMILY *VENERIDÆ*.

Meretrix casta Chemn.—A plain, heavily moulded, white porcelain *Meretrix*, with olive-ochraceous epidermis. A common Indian shell.

Meretrix castanea Lm. (= *M. morphina* Lm. var.) Several specimens. Singapore, Philippines.

Callista umbonella Lm.—Many specimens of this variable shell. Brazil (Cuming), Red Sea.

Crista divaricata Chemn.* Many specimens. Mozambique (Hanley), Red Sea, Philippines (Cuming).

Crista gibbia Lm.—One of our examples is unusually obese and large. Philippines, Red Sea.

Crista pectinata L.—Many, in all stages of growth. Indian Ocean, Philippines, Red Sea.

Circe personata Dh.—A delicate example of this somewhat common Indian species.

Sunetta meroë L. (=*S. picta* Schum.).—Two: a pretty species, and extremely variable in colour. Philippines.

Sunetta seminuda Rv.—One beautiful specimen, pale flesh coloured, with epidermis. Moluccas (Cuming).

Dosinia modesta Rv.—A few typical specimens. Spain (Paetel).

Dosinia radiata Rv. (=*Artemis amphidesmoides* Rv.).—Three or four, but only one exhibiting the characteristic rayed marking, so conspicuous in the figure in Conch. Icon. (pl. vii., fig. 37). The geographical distribution would appear to be wide, extending from the mouth of the R. Gambia, W. Africa, eastwards.

Dosinia salebrosa Römer.—A pure white shell, very delicately concentrically ribbed. Specimens in British Museum from Ceylon.

Chione graphica Lm. (=*Cytherea petechialis* Lm. var.).—Also reported from Sumatra and Japan.

Chione (Omphaloclathrum) gibbosula Dh.—A few examples, in good condition.

Chione (Omphaloclathrum) layardi Sow.*—Not, perhaps, quite typical. Found all round the coasts of India, Persian Gulf, and extending to the Andaman Islands. A very abundant Indian shell, occurring in every collection we have received from that country.

Chione (Omphaloclathrum) puerpera L.—Two; very large and fine, and well marked. Philippines.

Chione (Omphaloclathrum) scabra Hanl.—A minute, but elegantly chased *Chione*, of which there are some dozen examples in the collection. Philippines, New Holland.

Anaitis calophylla Hanl.—Two small but characteristic specimens.

Tapes (Textrix) malabarica Chemn.—Typical. Moluccas.

Tapes (Textrix) sulcosa Phil.—Remarkably fine, with the purple rays, and grey spots very distinct. Australia.

Tapes (Textrix) textrix Chemn.*—Some rather small specimens of ordinary character.

Tapes (Hemitapes) pinguis Chemn.*—A common Indian form. In the Römerian arrangement adopted in some museums this species is known as *Chione pinguis*. Ceylon, Tranquebar.

Irus macrophylla Dh. Quite young, no mature examples. A beautiful species, much lighter, more rounded, and having the laminæ more produced than our British *V. irus*; it also differs in being foliated and cancellated. Philippines.

Family *PETRICOLIDÆ*.

Petricola lithophaga Retr.—Indistinguishable from European specimens.

Family *DONACIDÆ*.

Donax (Hecuba) scortum L.*—Several, in various stages of growth. Cape of Good Hope.

Donax (Latona) abbreviatus Lm.*—Many examples. Philippines.

Donax (Latona) cuneatus L.—Four specimens. Ceylon.

Family *PSAMMOBIIDÆ*.

Gari anomala Dh.—One small but perfect example; also reported from Australia and New Holland.

Sanguinolaria hendersoni M. & S., *vide antea*, p. 33.[1]

Family *SOLENIDÆ*.

Siliqua radiata L.—Very beautiful and well-coloured examples. Sumatra.

Family *MESODESMATIDÆ*.

Mesodesma (Paphia) trigona Dh.—Large examples. Habitat hitherto unknown.

Family *MACTRIDÆ*.

Mactra attenuata Dh.—Several; somewhat attenuated posteriorly, and of a peculiar dark livid ash-violet, within and without. Habitat hitherto unrecorded.

Mactra fasciata Lm.—Several; agreeing with Reeve's description.

Mactra spengleri Born (=*M. lævis* Chemn. var.).—Our specimens are rather young, but agree with the description and figure in Reeve.

Schizodesma spengleri L.—Several typical examples.

Family *MYIDÆ*.

Corbula modesta Hinds*.—Five, all young, but agreeing with Bombay specimens received from Mr. Abercrombie. Philippines.

Corbula sulcata Lm.—Three young but recognisable specimens. Senegal.

Family *PHOLADIDÆ*.

Martesia striata L.—A drift species, its distribution therefore extending over the tropics. The "Challenger" specimens in the British Museum came from the Arafura Sea.

---

1 The description of this species was published on the cover of this Journal [illegible]

FAMILY *LUCINIDÆ.*

Lucina pisum Phil.—About eight examples of a highly chased and ribbed small shell, very rotund, evidently of wide distribution, as it has been dredged both at Port Essington, Australia, and Singapore.[1]

Lucina (Cyclas) semperiana Issel.—A minute *Lucina*, with some extraneous resemblance to the last (*L. pisum*). Like many of the genus, it is of wide distribution, being reported from Mazatlan (Paetel), and Gulf of Suez (Mus. Brit.)

Tellina (Tellinella) deltoidalis Lm.—Some very fine examples, quite typical in form. Australia, New Zealand.

Tellina (Tellinella) rostrata L.—Few, rather poor. Philippines.

Tellina (Tellinella) undulata Hanl.—Young examples only, but quite characteristic. West Columbia.

Tellina (Arcopagia) savignyi A. Ad.—Karachi (Townsend), and probably all round the Indian coasts. Red Sea (Paetel).

Tellina (Tellinula) valtonis Hanl.—Very thin and delicate, white, exhibiting a wonderfully beautiful iridescence.

Tellina (Tellinides) opalina Sow.—One specimen, agreeing with figure in Conch. Icon. Moluccas.

FAMILY *CLAVAGELLIDÆ.*

Brechites vaginifera Lm.—One, somewhat worn and broken, but evidently this well known Red Sea species.

1 Reeve, Conch. Icon., pl. xi., fig. 66.

NOTES ON A COLLECTION OF MARINE SHELLS FROM LIVELY ISLAND, FALKLANDS, WITH LIST OF SPECIES.

By J. COSMO MELVILL and R. STANDEN.

(Plate I., figs. 9-13; Plate II.).

(Read before the Society, Feb. 9th, 1898).

AMONGST the many recent additions to the collections of Mollusca in the Manchester Museum, a series of marine shells collected by Miss Cobb, at Shallow Bay, Lively Island, Falklands, is of particular interest. A few other species have also been added to Miss Cobb's collection, through the kindness of Mrs. Blake and Prof. D'Arcy Thompson. References to the molluscan fauna of this part of the world are somewhat meagre and scattered, so that the cataloguing of an authentic and characteristic collection like the present is not only an important local record, but also helps towards a better knowledge of the geographical distribution of certain species.

The archipelago of the Falkland Islands (Malvinas), forms a part of the "Magellanic Province" of Woodward,[1] which also includes the coasts of Tierra del Fuego, and the mainland of South America from Port Melo, on the east coast, to Concepcion, on the west. They are situated in Lat. 51° 30' S., and cover a space of 120 by 60 geographical miles, or little more than half the size of Ireland. They are a treeless expanse of moorland and bog, and bare and barren rock, and their wild and rugged shores are washed by tempestuous seas, swarming with mollusca and other forms of marine life, which find ample shelter and sustenance amongst the dense masses of "kelp"—a giant seaweed (*Macrocystis pyrifera*) growing in profusion on every tidal rock, and forming a barrier to the terrible breakers of the western ocean, which no mass of rock not thus protected could long withstand. Woodward assigns 45 species of mollusca to the "province" in general, but gives only the following as being known from the Falklands :—

Scalaria brevis	*Trochita pileolus*
Margarita malvinæ	*Astarte longirostris*
Fissurella radiosa	*Cyamium antarcticum*
Patella barbara	*Modiolarca trapezina*
P. zebrina	*Cardita thouarsii*
P. deaurata	*Venus exalbida*
Scissurella conica	*Lyonsia malvinensis*

He further remarks that "eleven of these have not been met with elsewhere."

1 Manual of Conchology, 1880, p. 52.

Dr. Paul Fischer[1] enumerates 81 species as inhabitants of the "Province magellanique," and to Woodward's list of Falkland shells he adds—*Euthria antarctica, Trochus tæniatus, Fissurella picta, Puncturella conica, P. falklandica* and *Modiolarca pusilla*.

The 'Challenger' expedition collected 18 species at the Falklands. Of these, *Calyptræa pileus, Photinula cærulescens, Trophon liratus*, and possibly two unidentified species of *Calyptræa* and *Lamellaria* included in the list, are not represented in the Cobb collection. The 'Challenger' specimens were mostly obtained from the "kelp" in from 5 to 12 fathoms. Miss Cobb's shells were all collected on the beach, but in the majority of cases are in very good condition, and show but slight signs of sea-wear. To several of the shells of *Mytilus* and *Patella* some most beautiful specimens of *Microporella ciliata, Cribrilina labiosa, C. monoceros, Mucronella tricuspis, Cellepora tiara, C. punctulata* and other Bryozoa are attached. We are indebted for these identifications to Miss E. M. Pratt, by whom they have been carefully studied, in connection with another collection of zoological specimens received by the Museum from Mrs Blake, of Hill Cove, Falklands.[2] Included in Mrs. Blake's collection are also a number of specimens of *Trophon muriciformis, Euthria fuscata, Photinula violacea* and *Fissurella mexicana*, with the animals preserved in spirit, together with several Brachiopods, of which two species, viz.:—*Terebratella magellanica* and *T. dorsata*, likewise occur in the Cobb collection.

LIST OF SPECIES.

Siphonaria lessoni Blainv.—Several typical examples. Recorded also from Chili (Paetel).

Lachesis euthrioides sp. n. (Pl. I., fig. 9).

L. testa fusiformi, tenui, læté castaneo-brunnea, anfractibus septem, quorum duobus albatis, apicalibus, cæteris apud suturas impressis, tumidulis, longitudinaliter crassicostulatis, costis ad juncturas costularum spiralium nodulosis, nodulis lævibus, nitidissimis; apertura arcto-rotunda, labro convexo, tenui, canali brevi, curta, margine columellari paullum excavato. Long. 8, *lat.* 3 *mm.*

Fusoid superficially, this little cancellated shell is undoubtedly a *Lachesis*. It is of a bright chestnut brown, somewhat shining, fusiform, seven-whorled, two whorls apical and white, the rest impressed at the sutures, tumid, longitudinally thickly costate, and transversely filo-costulate, shining, noduled at the points of junction. The aperture is ovate-rotund, outer lip thin, canal short, columellar margin slightly excavate. Three specimens.

---

1 Manuel de Conchyliologie, 1887, p. 172.

2 For list of these forms with notes, see *Manchester Memoirs*, vol. 42, 1898.

The genus *Lachesis* Risso, as restricted, now embraces about thirteen species, inclusive of a new species (*L. bicolor* Melv.) from the Arabian Sea.[1] They are mostly extra-tropical, indeed, the type *L. minima* Montagu is a well known inhabitant of our southern British coasts. Besides this, three are Mediterranean, one Japanese, one from the Island of St. Paul, one (*L. sulcata* Hutton) from New Zealand, and another (*L. meridionalis* E. Sm.) from the Strait of Magellan. This is a curious species, the longitudinal ribs being quite obsolete on the lower half of the body-whorl; while, at the periphery, there is one series of revolving tubercles, and the size is only 4 mm.

Voluta (Cymbiola) ancilla Soland.—One very large dead specimen. Gould gives a capital figure of the animal, but, like Sowerby and Kiener, describes this species as *V. magellanica* (*non* Lm.). It is *V. gracilis* Wood. D'Orbigny records it from Strait of Magellan, and Paetel from coasts of Patagonia.

V. (C.) becki Brod.—Two; one quite juvenile (Pl. I., fig. 11); the other a full-grown specimen, measuring a little over nine inches in length, with, presumably, about an inch of apex missing. It is yellowish chestnut in colour, with longitudinally undulated streaks. In the Philadelphia Academy of Natural Sciences there is a specimen measuring 14 inches in length (Tryon). Hitherto the precise habitat of this large species appears to have been unknown, though generally assigned to Patagonia. Its occurrence in this collection is, therefore, of particular interest.

V. (C.) magellanica Lm.—One small specimen. Patagonia (Paetel).

Euthria antarctica Rv.—Two, quite typical.

E. fuscata Brug.—This species was described originally by Bruguière as a *Buccinum*, and recorded as occurring abundantly on the coast of Peru. In Mrs. Blake's collection there are many examples, in all stages, while in the Cobb collection there are but two, both typical.

E. plumbea Phil.—One rather young example of what appears to be this species, which occurs from Cape Horn to Chili, also Japan. Its synonyms are *Fusus rufus* Homb. & Jacq., *Buccinum magellanicum* Phil., *B. patagonicum* Phil., and, probably, *Euthria ferrea* Rv., and *E. viridula* Dkr.

Trophon albolabratus E. Sm.—One, quite typical, and in good condition. Although Tryon unites this with the very variable *T. geversianus* Pallas, he seems to have been somewhat hasty in his decision, for he admits that he has never seen the species, which Mr.

1 *Manchester Memoirs*, vol. 42, no. 4, p. 14, pl. 1, fig. 17, 1898.

E. A. Smith described from Kerguelen. It is a much more solid shell than *T. geversianus*, and narrower, with fewer and less regular spirals; the suture is not so deeply impressed; the lip is thicker, and the mouth less oblique, whilst the canal is much shorter. Specimens were dredged by the 'Challenger' Expedition in three stations off Royal Sound, Kerguelen, from volcanic mud, in 25 to 60 fathoms (Watson).

T. buccineus Gray.—One specimen, in perfect condition, which exactly matches the excellent figure copied by Tryon. No description or locality is given, but the shell is allied to *T. plumbeus* A. Gd., and is a light chocolate-brown colour, deeper within the aperture.

T. cretaceus Rv.—One rather large example, in the usual worn condition. Recorded from the coast of Chili.

T. geversianus Pall. (Pl. I., fig. 10, Pl. II.).—Six examples of this beautiful species. One is a particularly handsome specimen, snowy white, with well developed broad frill-like lamellæ, and in excellent condition. It indeed so far surpasses in size and beauty the types figured in Reeve's "Conchologia Iconica" and elsewhere, that it has been thought worth while to figure two aspects of this shell. This particular specimen is in J. C. Melvill's collection. Two others almost equal it in size, but are not so perfect in condition. The rest are juvenile. With the collection there is a string of egg-capsules of this species. They are of a yellowish colour, spoon-shaped, closely set together, and in size average 20 × 12 mm. In each capsule there are a quantity of embryos, measuring 2·5 mm. in length; the contents of two capsules were carefully counted and found to contain, respectively, 74 and 112 baby shells. One of these is figured (Pl. I., fig. 10). The species is extremely variable, and occurs from Magellan's Strait to Chili. The synonymy is extensive and includes, amongst others, *Buccinum fimbriatum* Martyn, *Murex magellanicus* Gm., *M. foliatus* Schum., *M. peruvianus* Enc. Méth., *M. lamellosus* Dillw., and *M. patagonicus* D'Orb.

T. laciniatus Martyn.—One, not quite adult, but otherwise very perfect. The absence of the characteristic latticed sculpture in the interstices between the lamellæ readily distinguish this from *T. geversianus*. It is recorded from Magellan's Strait by Tryon, and from the Chonos Archipelago by Reeve. Tryon unites this species and *T. antarcticus* Phil., and expresses some doubt as to whether these South American forms really belong to the genus *Trophon*.

T. muriciformis King.—Five very beautiful and perfect examples. An ovately fusiform, cinereous species, with tumid, cancellated whorls; dark chestnut aperture, and crenulated lip, which amply differentiates it from *T. geversianus*, with which Tryon is inclined to group it. There are also two specimens in Mrs. Blake's collection. Recorded from Strait of Magellan.

Monoceros calcar Martyn.—One example of this extremely variable shell, of the form with exserted spire to which Lamarck gave the name of *M. imbricatus*. The species occurs from Cape Horn to Chili, where it attains its maximum heaviness of shell. Between this massive form (*M. crassilabrum* Lam.) and the type there ranges a long series of transitional forms, the more marked of which have severally been named *M. striatum* Lm., *M. glabratum* Lm., *M. globulus* Sow., *M. costatum* Sow., *M. citrinum* Sow., and *M. acuminatum* Sow.

Crepidula dilatata Lm.—A good series. One specimen is the form described as *C. pallida* by Broderip; the rest are fairly typical. The type is a large, rugose, inflated form, having the apex much curved to the side of the shell, and the inner margin usually deeply stained, or radiately lineated with chestnut. It has many synonyms, *e.g.*, *C. depressa* Dh., *C. peruviana* Lm., *C. patula* Dh., *C. arenata* Brod., *C. adolphei* Less., *C. grandis* Midd., *C. princeps* Conr., *C. nummaria* A. Gd., etc. It is generally distributed all along the western coast of America, from Patagonia to Alaska, and also Kamchatka.

Calyptrea (Trochita) radians Lm.—Three specimens, all of them identical with *C. corrugata* Rv. Other synonyms are *C. peruviana* Dh., *C. concamerata* Martini, *C. costellaria* Phil., *C. trochiformis* Gm., *C. trochoides* Dillw., *C. araucana* Less., *C. sordida* Brod., and *C. spirata* Forb. It is recorded from Chili and Peru, also from the Island of Inagua, Bahamas (Tryon).

Natica impervia Phil.—Two typical specimens. In this species the callus completely fills the umbilicus. Magellan's Strait (Tryon).

N. magellanica Phil.—Three. Tryon appears to think this may be a form of *N. patagonica* Phil., from Strait of Magellan.

Scalaria (Opalia) magellanica Phil.—Two typical specimens. Also sent by Mrs. Blake.

Photinula expansa Sow.—Three specimens, all of the typical light olive colour, with green and iridescent interior. It also occurs at Kerguelen Island, Strait of Magellan, and South Georgia. The principal synonyms are *Margarita expansa* Sow., *Trochus expansus* Phil., *Photina expansa* A. Ad., *Margarita hilli* Forbes, etc.

P. tæniata Wood.—Several typical examples of this, the largest and most striking species in the small group of *Margarita*-like shells found only in Antarctic America, for which H. & A. Adams instituted the genus *Photinula*. It is imperforate, with a brilliantly nacreous interior, and easily recognizable by the spiral pink bands and lines on a shining white ground. The synonymy is extensive, and includes *Margarita tæniata* Sow., *Trochus bicolor* Less., *T. lineatus* Phil. (*non* da Costa), *T. hombroni* P. Fisch., *T. purpuratus* Forbes, *Photina cærulea*, A. Ad., etc. It occurs also in Magellan's Strait (Tryon).

P. violacea King.—Seven specimens. A more conical shell than *P. expansa*, with a smaller aperture, and of a purplish-pink colour. Its synonyms are *Margarita persica* A. Gd., *M. violacea* King, *M. magellanica* H. J., *Trochus violaceus* Phil., and probably *Margarita magellanica* A. Gd., but Tryon is inclined to doubt the identity of Gould's species with *P. violacea*. Recorded from Strait of Magellan and Orange Harbour, Tierra del Fuego.

Fissurella darwini Rv.—One dead, but characteristic example of a form which appears to occupy an intermediate position between *F. picta* Gm. and *F. maxima* Sow. It is more conical than the latter, and less elevated than the former, and is sculptured with numerous radiating ribs, which are low and uneven, but scarcely to be called granose. The inside is white, with a blueish-black border. Recorded from Strait of Magellan (Tryon).

F. picta Gm.—Two full grown and typical examples of this handsome shell. Its recorded habitats are Strait of Magellan (D'Avila), Chili and Valparaiso (D'Orbigny).

F. polygona Sow.—Several typical specimens of this most beautiful shell, which is distinguished by its scabrous striæ and chain-like ribs, usually rayed with purple. The inside is white, and the margin spotted. None of the shells in this collection are quite as large as the one figured by Sowerby, and one specimen is without colour, though in the finest possible condition.

F. mexicana Sow.—Four specimens. Tryon considers that the locality given by Sowerby and Reeve ("Real Llejos, Mexico") needs confirmation, and hints that it may be Sowerby's *F. oriens*; but the shells now before us are well distinguished from that species by the finely decussated, close-set radiating grooves, and white-bordered orifice.

Puncturella (Cemoria) falklandica A. Ad.—One typical example. Dall is doubtful whether this is separable from the British *P. noachina*, from which, however, it differs somewhat in its shorter fissure and septum, more distant and equal ribs, and less posterior apex. Recorded from west coast of Patagonia, in 449 faths. (Tryon).

Patella ænea Martyn.—A large and representative series of all the principal varietal types acknowledged by Tryon in (*a*), typical *P. ænea*; (*b*), var. *deaurata* Gm.; (*c*), var. *magellanica* Gm. The type is a very solid shell, elevated, with strong radiating ribs, brownish ash-coloured exteriorly, with a lustrous bronze-tinted interior, and scalloped margin. *P. gaudichaudi* Blainv. is a synonym. The var. *deaurata* is chiefly distinguished from the type by the thin, oblong, depressed shell, showing dark-brown radiating stripes interiorly, and the apex curving forward. *Patella cymbularia* Del., *P. ferruginea* Sow., *P. delesserti* Phil., *P. varicosa* Rv., and *Nacella strigatella* Roch. & Mab., are also synonyms. Tryon also considers that *P. polaris* Marts. & Pfr. is merely a form of *P. deaurata*, and that *P. varicosa* Rv. is identical with Woods' & Gmelin's *P. flammea*. Var. *magellanica* Gm. is of a rounder form than typical *P. aenea*, and has a more central, erect, and elevated apex. Variations of this form have been described by Reeve, under the names of *P. atramentosa*, *P. venosa*, and *P. chiloënsis*, whilst Rochbr. & Mab. coin many species from it. Both typical *P. aenea* and the vars. *deaurata* and *magellanica* have hitherto only been recorded from Strait of Magellan.

Acmæa textilis A. Gd.—Six specimens. This species seems to possess good and distinctive specific characteristics, although included by Tryon in the synonymy of *A. persona* Eschz., an excessively variable species, which ranges from Sitka to Turtle Bay, Lower California.

Scurria scurra (Less.).—A very fine example of this has been added to the collection through the kindness of Prof. D'Arcy Thompson. It was collected by Mr. J. Cooper, in 1893, and is of the somewhat depressed terraced form figured by Tryon.[1] The range of this species is given by Tryon as extending from 12° to 41° S. Lat., west coast of South America. Its synonyms are *Patella scurra* Less., *Acmæa scurra* D'Orb., *Lottia pallida* Sow., *Lottia conica* A. Gd., and *Acmæa cymbula* Hupé.

Nacella cymbularia Lm.—Three specimens, of a pale horn colour, with silvery iridescent interior. Tryon restricts *Nacella* to one species: *N. mytilina* Helbl., which lives upon the great seaweeds of the Tierra del Fuego shores, and he includes *N. cymbularia* amongst its numerous synonyms.

Chiton (Plaxiphora) setiger King.—One typical example of this well-known South American species.

Lima falklandica A. Ad.—One perfect specimen, together with odd valves of a very delicate shell with some affinity to *L. loscombi* Sow.

Pecten (Pallium) corneus Sow.—Valves only. The typical form. Recorded from Strait of Magellan.

P. rufiradiatus Rv.—Many valves, distinctively representative of the species. It is recorded from Strait of Magellan by Reeve, and is a characteristic form of the type of *Pecten* peculiar to that locality;

1 Man. Conch., vol. 13, pl. 39, figs. 26-27.

but it differs from the more orbicular *P. patagonicus* King, in the camparatively greater height of the shell, crenulated ribs, and unequal ears.

Mytilus bifurcatus Conr.—A peculiar looking shell, about 25 mm. in length, strongly grooved, triangularly ovate, very gibbous, and of a dark indigo blue-black colour. Reeve does not give any locality, but it is recorded from California by Paetel.

M. magellanicus Chemn.—Three young and two adult examples of this fine pear-shaped shell. It is blue-black in colour, with waved crenated ribs. The larger shells bear beautiful growths of Bryozoa, etc.

M. ungulatus L.—With the exception of size, there seems little to separate this from our *Mytilus edulis.* It seems to run through the same gradations of form, including typical var. *gallo-provincialis* Lm. Recorded from Chili (Cuming). Paetel makes *M. ungulatus* synonymous with *M. latus* Lm. from New Zealand, but this can scarcely be correct.

Modiolarca pusilla A. Gd.—One typical example. Recorded from Kerguelen (Fischer).

M. trapezina Lm.—Nine specimens, in various stages of growth, of a very elegant shell, varying in colour from bright orange to purplish yellow or rich sienna-brown. Recorded from Strait of Magellan (Paetel); Kerguelen and Auckland (Fischer).

Cyamium falklandicum sp. n. (Pl. I., fig. 12).

C. testa mediocri, æquivalvi, inæquilaterali, oblongo-rhomboidali, tenui, albida; valvis sub lente undique concentricé striatis, striis rudibus, posticum apud marginem rugoso-laminatis, umbonibus prominulis, contiguis; valvis dorsaliter posticé leniter declivibus, prolongatis, anticé rotundatis, ligamento corneo, externo, valvam apud rectam dente cardinali magno, bifido, apud sinistram duobus minoribus; superficie interna alba, parum nitente, linea palliali indistincta, paullum sinuosa. Alt. 7, *lat.* 12, *diam.* 4 *mm.*

This interesting addition to a very circumscribed genus has been confused with *C. antarcticum* Phil., by Gwyn Jeffreys.[1] This latter, however, is quite distinct, and correctly-named examples exist in the British Museum, where also is this species without a name. It is smallish in size, equivalve, very inequilateral, rhomboidal, thin, white, the valves concentrically striate, and towards the posterior margin wrinkled-laminate; the ligament is horny and external, the posterior dorsal margin gradually sloping, prolonged, anteriorly rounded, the cardinal tooth in the right valve is large and bifid, in the left there are two smaller teeth; within the surface is

1 Brit. Conch., vol. 2, p. 257.

white, scarcely shining, the pallial line to some extent sinuous, but not very distinct.

Chione (Omphaloclathrum) exalbida Chemn.—Three examples, varying in growth and development. A massive, flat, oblong shell, with raised concentric striæ. Occurs in Straits of Magellan.

Cryptodon falklandica E. Sm.—A shell with some affinity to *Axinus flexuosus* Mont.

Saxicava antarctica Phil.—A stout coarse shell, found burrowing in the roots of the large seaweed. Recorded from coasts of Chili (Paetel).

Thracia antarctica sp. n. (Pl. I., figs. 13, 13a).

T. testa deformi, ferè æquivalvi, sordidè alba, tenui; valvis ambabus concentricè rudistriatis, convexis, utrimque hiulcis, anticè rotundatis, posticè truncatis, brunneo-sordescentibus, rudilamellosis; umbonibus in uno specimine prominulis, incurvis, contiguis, in altero ferè immersis; margine dorsali posticè paullum excavato, anticè leniter declivi, ventrali ferè recto; dente cardinali magno, cochleari; superficie interna alba, paullum prismatica, linea palliali obscura, sinuosa. Lat. 15, *alt.* 11, *diam.* 6 *mm.*

To some extent resembling *T. distorta* Phil. from North Europe, or *T. cuneolus* Rv., this very interesting little form differs from both in decidedly less rotundity of outline; it is, indeed, a far more typical *eu-Thracia*, though to some extent liable to the *Saxicava*-like deformity so often present in our North European *Th. distorta.*

It is of a dirty white, posteriorly stained with brown, and truncate, anteriorly rounded, the umbones contiguous and prominent in one specimen, while in another they are almost immersed; the dorsal margin is posteriorly slightly excavate, anteriorly gently sloping, the ventral margin almost straight. The surface is concentrically rudely striated, the valves convex, almost equal. Interior surface slightly nacreous, whitish, pallial line obscure, sinuous.

Only one *Thracia* (*T. similis* Conr.), of quite a distinct group, has, till now, been reported from South American shores.

ON LATIRUS ARMATUS Ad.

By J. COSMO MELVILL.

(Read before the Society, March 9th, 1898).

Miss Edith C. Wilson has presented a small collection of marine mollusca, gathered by herself in the Canary Isles, to the Manchester Museum, Owens College. The majority of the shells call for no special remark, but amongst them is one dead though perfect and well-developed specimen of *Latirus armatus* Ad., entirely free from any nullipore or other extraneous marine growth, and consequently in a perfect condition, so far as the shell is concerned, for investigation.

It is a great pity we know so little about the animal. The only item of information vouchsafed us, so far as I can find, is that it is of a red colour. That, of course, is a distinctive attribute of all *Latiri* proper, but we cannot help hoping that full anatomical details of this much discussed and variable form may be forthcoming at no distant date. In the meantime the following is the history of *Latirus armatus* Ad. In 1838 Dr. Gray described a form as *Turbinella spinosa*, which is in all probability this species. The name *T. spinosa* Martyn being already in use, reduced Gray's name to a synonym, and in 1854[1] Mr. A. Adams described from the Cumingian cabinets eleven *Latiri*, without figures or information as to size and in exceedingly bald and bare phraseology. Amongst these we find:—"LATIRUS ARMATUS A. Adams. *L. testa ovato-fusiformi, umbilicata, spira apertura breviore, fulva, epidermide fusca obtecta; anfractibus longitudinaliter plicatis, lira prominenti transversa (muricata ad plicas) in medio anfractuum ornatis, ultima liris minutis instructo; apertura ovali, canali recto, aperto, columella obsolete plicata, plicis quinque, labro intus sulcato, margine crenato.*

Hab.: California (Mus. Cuming).

This is an ovately fusiform shell, with a muricated transverse ridge in the middle of the whorls, which are covered with a brown epidermis."

Ten other *Latiri* were described at the same time.

The one example of the Cumingian collection was in 1866 transferred to the British Museum, and lay neglected and unobserved for some years till in 1873 the Rev. R. Boog Watson received from Madeira an extraordinary shell with large umbilicus and consequent pseudo-distortion of mouth and canal which, acting on advice tendered him by Dr. Gwyn Jeffreys, F.R.S., and Dr. Paul Fischer, he raised to the rank of a new generic type, under the name *Chascax maderensis* Watson.[2] His description is minute to a nicety and exact

1 *Proc. Zool. Soc.*, 1854, p. 314.
2 *Proc. Zool Soc.*, 1873, p. 361.

in every particular. There is only room to transcribe the first opening sentences :—

"*Chascax* gen. nov. Watson. Shell spindle-shaped, strongly umbilicated, longitudinally ribbed and spirally ridged, but without varices. Epidermis horny. Mouth edge angulated. Outer and inner lip quite smooth. Canal long, narrow, and deep, bent a little to the left, but not at all reversed in front. Operculum strong, horny; nucleus terminal, internally strengthened by a broad ridge all along the right margin."

In 1886, the "Report on the Gasteropoda collected by H.M.S. 'Challenger'" was published, and the Rev. R. Boog Watson names as *Latirus armatus* Ad. the single specimen dredged on this expedition at Station 7 P. Lat. 28° 35' N., Long. 16° 5' W., off Teneriffe, on volcanic sand, 10th February, 1873.[1] If we refer to his preliminary paper[2] on the same subject, we find the shell named *Fasciolaria maderensis* n. sp., and referred to the *Turbinella carinifera* auct., non Lm. Mr. Watson also expresses a doubt whether his *Chascax maderensis*, referred to above, may not be a very aberrant variety. These remarks he repeats almost in extenso in the revised account,[3] and I think an examination of the *Chascax* in the British Museum will prove that his doubt was well founded. Indeed the *Chascax maderensis* is to the typical *L. armatus* exactly as the widely umbilicated form of *Latirus undatus* or *L. infundibulum* is to the less developed shells. It is the tendency of typical *Latiri* to form shells with this (no doubt more or less monstrous) characteristic. The umbilicus is deep seated, and in Miss Wilson's specimen, which is intermediate between the abnormal *Chascax* and the moderate scarcely umbilicate *L. armatus* Ad.,[4] the narrowness is remarkable.

The specimen before us is of pale buff colour, decorticated, heavy, seven whorled, upper whorl angulated in the middle, the upper portion sloping to the suture, the lower straight; the median angulation is sharply noduled, the lower whorl sloping from the suture for about one-fifth of its surface, then transversely angulated and conspicuously sharply noduled; below this a median portion runs nearly straight. Longitudinally, once very lightly transversely lirate, followed by two stronger spiral-raised somewhat noduled ribs. Towards the base are two more light spiral costæ, the aperture is ovate, narrowed peculiarly, as if distorted by the umbilical extension, towards the canal; the outer lip is five or six times grooved; columellar plaits almost if not quite obsolete; umbilicus narrow, but pronounced and deep; operculum not present.

---

1 'Challenger' Gasteropoda, p. 243.
2 *J. Linn. Soc.*, vol. 16, p. 336, 1883.
3 *Op. cit.*, p. 244.
4 Compare 'Challenger' Gasteropoda, pl. 13, fig. 1.

The true *Latirus cariniferus* Lm. from the Pacific is quite distinct. In this the shell slopes away immediately below the very prominent median spiral angulation at the last whorl. Indeed the abundant *L. polygonus* will remind one more of the typical *L. armatus*, but the whole texture being so different, no one could for an instant suppose that they were identical.

California, given by Adams as the habitat of the Cumingian type, must be erroneous. I see no reason for altering the sequence of this species in the catalogue of *Latirus* and *Peristernia*, with *L. distinctus*, *L. cariniferus*, and *L. polygonus* as nearest allies.

[PAGE 95

Note on Terebra eximia Dh.—On looking through a miscellaneous series of beach-collected marine shells from Borneo, received from the late Rev. W. Turner, I was highly gratified to recognize amongst some other *Terebræ* a good specimen of *T. eximia* Dh. It is 36 mm. in length, and although rather smaller than the one recorded by Mr. J. C. Melvill and myself, in our recent paper on "Madras Mollusca" (*antea* p. 35, Pl. I., fig. 8), it is equally good as regards condition, sculpture, and marking. The discovery of another example of this rare and beautiful species so soon after our Madras record, is extremely interesting, especially as the type specimen, from an unknown locality, was described so long ago as 1859 (*Proc. Zool. Soc.*, 1859, p. 314), and has hitherto remained unique.—R. STANDEN (*Read before the Society*, Feb. 9th, 1898).

[PAGE 85

Note on Cypræa rashleighana.—The above Cowry was described in 1887,[2] and in the following year was re-figured, the original description being repeated in the "Survey of the Genus *Cypræa*, 1888."[3] Although the habitat was queried it seems probable that the type came from the neighbourhood of Hongkong. Since this time three or four specimens have occurred amongst the Hadfield Mollusca from Lifu; these, however, are either too young or in a not very satisfactory state of preservation. My object in alluding to this species at the present opportunity is to call attention to a very beautiful and large example which has been for years in the National Collection at South Kensington, having formed part of the Cumingian stores. This was figured by Mr. Lovell Reeve[4] as a stunted form of *C. tabescens* L., but has been overlooked by Sowerby[5] and by Mr. Raymond Roberts in the "Monograph of *Cypræa*."[6] Rather blindly following Reeve in 1888,[7] I signalised this as var. *a* of *C. tabescens* under the proposed varietal name of *latior*. Mr. Edgar Smith being disposed to allow it specific rank, labelled it in the National Collection "*latior* Melv." Last year, however, it was closely examined by us both, in comparison with the original type of *C. rashleighana*, and pronounced identical. The pyriform shape, different dentition, narrower aperture, small clearly defined dark-brown lateral punctuation, with other characteristics, differentiate this species from its allies, *C. tabescens*, *C. teres* and *C. interrupta*.—JAMES COSMO MELVILL (*Read before the Society*, April 13th, 1898).

---

2 *J. Conch.*, vol. 5, p. 288. 3 *Manch. Mem.* (4), vol. 1, p. 218, 219. 4 Conch. Icon., pl. 14, no. 66a, 1845. 5 Thes. Conch. 6 Tryon, Man. Conch., vol. 7, 1885. 7 *Loc. cit.*, p. 218.

EXPLANATION OF PLATE I.

Fig. 1.—*Cerithium carnaticum* sp. nov. page 31
,, 2.—*Colina selecta* sp. nov. ... ,, 31
,, 3.—*Rissoina (Morchiella) thaumasia* sp. nov.... ,, 31
,, 4.—*Syrnola maderaspatana* sp. nov. ,, 32
,, 5.—*Turbonilla coromandelica* sp. nov. ,, 32
,, 6.—*Cadulus anguidens* sp. nov. ... ,, 32
,, 7.—*Sanguinolaria hendersoni* sp. nov. ,, 33
,, 8.—*Terebra (Euterebra) eximia* Dh. ,, 34
,, 9.—*Lachesis euthrioides* sp. nov. ... ,, 98
,, 10.—*Trophon geversianus* Pall. (juv.) ... ,, 100
,, 11.—*Voluta (Cymbiola) becki* Brod. (juv.) ,, 99
,, 12.—*Cyamium falklandicum* sp. nov. ,, 104
,, 13, 13a.—*Thracia antarctica* sp. nov. ,, 105

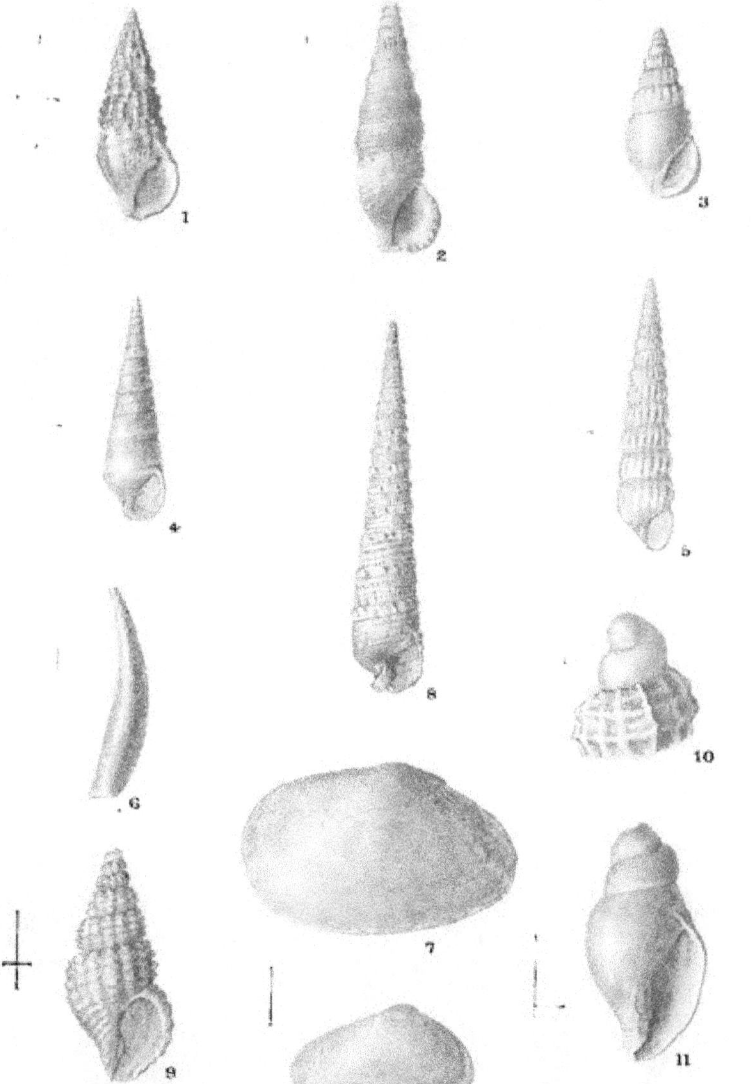

G. M. Woodward del. et lith.

West, Newman imp.

MADRAS & FALKLAND ISLANDS MOLLUSCA.

EXPLANATION OF PLATE II.

Trophon geversianus Pall., natural size.

(See page 100).

From a photograph by Mr. Edward Ward. The original specimen is now in the collection of Mr. J. Cosmo Melvill.

TROPHON GEVERSIANUS.

MUSEUM REGULATIONS.

1.—The Museum is open to the public every week day from 11 a.m. to 5 p.m., on Sundays from 2-30 to 4-30 p.m., and on the first Wednesday in each month from 7 to 9 in the evening. Admission free and without ticket.

2.—The Museum and Museum Library are further open to persons desirous to make use of them for the purposes of study. The Keeper will issue tickets of admission on suitable recommendation.

3.—Students of the College are admitted to the Museum on any day in the week between the hours of 10 a.m. and 5 p.m. Those whose studies necessitate access to the specimens and the loan, for use in the Museum buildings only, of the books in the Museum Library, may procure Students' tickets from the Keeper, on the recommendation of the Professors or Lecturers in the several departments, such tickets to be valid only for the session in which they were issued.

4.—The Museum is closed on Good Friday and Christmas Day.

www.ingramcontent.com/pod-product-compliance
Lightning Source LLC
Chambersburg PA
CBHW022017190326
41519CB00010B/1549
* 9 7 8 3 7 4 4 7 2 9 8 8 8 *